GAS-INSULATED TRANSMISSION LINES (GIL)

GAS-INSULATED TRANSMISSION LINES (GIL)

Hermann Koch

Energy Transmission, Siemens AG, Germany

IEEE PRESS

A John Wiley & Sons, Ltd., Publication

This edition first published 2012
© 2012 John Wiley & Sons, Ltd

Registered office
John Wiley & Sons Ltd, The Atrium, Southern Gate, Chichester, West Sussex, PO19 8SQ, United Kingdom

For details of our global editorial offices, for customer services and for information about how to apply for
permission to reuse the copyright material in this book please see our website at www.wiley.com.

Library of Congress Cataloging-in-Publication Data

Koch, Hermann
  Gas insulated transmission lines (GIL) / Hermann Koch.
    p. cm.
  Includes bibliographical references and index.
  ISBN 978-0-470-66533-6 (cloth)
  1. Electric cables–Gas insulation. I. Title.
  TK3331K63 2011
  621.319′34–dc23

                        2011024794

A catalogue record for this book is available from the British Library.

Print ISBN: 9780470665336
ePDF ISBN: 9781119953074
oBook ISBN: 9781119953081
ePub ISBN: 9781119954354
Mobi ISBN: 9781119954361

Typeset in 10/12pt Times by Aptara Inc., New Delhi, India
Printed and bound in Singapore by Markono Print Media Pte Ltd

Contents

Foreword

Environmental constrains have intensified the interest in underground transmission lines as an alternative solution to overhead lines. Nowadays, different options of underground transmission lines are available. The most common solution are cables, however, gas-insulated lines (GIL) are also applied as an alternative. GIL can be laid directly buried, in a passable or non-passable tunnel or on a gantry. In particular, the directly buried type of GIL is attractive because it can be verified over distances of some kilometres up to some tens of kilometres by adjustment to the landscape and to impediments. GIL solutions offer different advantages, such as high current-carrying capacity, low losses, low charging capacity etc.

Although GIL technology is mainly based on the experience with gas-insulated switchgear, a lot of specific features and criteria in design, installation and maintenance have to be considered. The present book written by Hermann Koch – who has been involved in this subject for nearly two decades – gives a broad overview on all issues of this technology, which was introduced in the 1970s and has since been applied more and more as an alternative to cable solutions. It differs between the first generation of GIL, filled with pure SF_6 and consisting of flanged sections, and the second generation of GIL, filled with a gas mixture (SF_6 and nitrogen) and comprising welded enclosures and conductors. After presentation of the dielectric properties and design features, the assembly and laying procedure is described and the quality assurance measures and testing methods – including diagnostic tools – are given. At this stage the special requirements of the different laying options are particularly taken into account. Furthermore, system and network issues, the environmental impact and economic aspects are considered. Finally, various worldwide GIL applications are illustrated and a comparison with other transmission systems regarding technical and site features as well as economical aspects and soft parameters like aesthetics, non-visibility and noises is presented.

This book will be invaluable for engineers involved in the special requirements of today's system layout. It should be of particular interest to students specializing in power system engineering, but also contains fruitful information for readers who want to get a general overview of the different underground transmission technologies.

Professor Claus Neumann
Essen, Germany

Acknowledgements

This book reflects the information and experience gleaned from the development, manufacture, assembly and operation of gas-insulated transmission lines over the last 20 years. Contributions to this book have come from many sides, and many experts in the field have been involved.

My thanks go to the IEEE (who asked me in the first place to write a book on gas-insulated transmission lines) and to the IEEE PES Substations Committee, which supported me.

The material in this book is based on, and derived from, more than 130 publications – most of them written by the author together with many co-authors. Not everyone can be mentioned here who participated in the development of the second generation of gas-insulated transmission lines, which is the core content of this book. I would like to give thanks to those experts who have contributed in joint or own publications and with many persona discussions about gas-insulated technology:

Abilgaard, Max; Accourt, Christian; Alter, Joachim; Ammann, M.; Arora, Arun; Aschendorf, M.; Bär, G.; Becks, Martinus; Benato, Roberto; Boeck, Wolfram; Bolin, Phil; Bowman, Gary; Brachmann, Patrick; Boisseau, Christoph; Brückner, Erhard; Buchholz, Bernd; Chakir, Abdellah; Chakravorti, Sivaji; Colombo, Enrique; Connor, Theodor; Cousin, Vincent; Degen, Wolfgang; Dießner, Armin; Di Mario, Claudio; Drews, Anja; Dürschner, Rolf; Ebner, Andreas; Engelhardt, Gerhard; Feldmann, Dominique; Finkel, M.; Fitzgerald, Patrick; Gatzka, Uwe; Glaubitz, Peter; Glaubrecht, Arnd; Goto, Kiyoshi; Gorablenkow, Jörg; Graf, R.; Grund, Armin; Harnass, Olaf; Hermann, Frank; Hillers, Thomas; Hinrichsen, Volker; Hennigsen, Claus; Hofmann, Lutz; Hopkins, Mel; Hühnerbein, Benjamin; Ikeda, Hisatoshi; Imamovic, Denis; Jesberger, Michael; Kaul, Guido; Kelch, Thomas; Kieper, Mario; Kindersberger, Josef; Kobayshi, Shinichi; Köpke, Kathi; Kunze, Dirk; Kynast, Edelhard; Lampe, Karl-Heinz; Lausegger, Markus; Le, Manh-Hung; Matsumura, Motofumi; Meinherz, Manfred; Monard, Denis; Muhr, Michael; Neumann, Claus; Obst, Dietmar; Olson, Erhard; Oswald, B. Pfeiffer, Wolfgang; Pigini, Alberto; Plath, Ronald; Pöhler, Stefan; Polster, Klaus; Povh, Dusan; Rathke, Christian; Renaud, Francois; Retzmann, Dietmar; Rieder, Ludwig; Ruan, Quanrong; Sabot, Alain; Schedl, Stefan; Schichler, Uwe; Schöffner, Günther; Schramm, Heinz; Schreieder, Alfons; Schütte, Andreas; Seiter, Egon; Sieber, Peter; Siebert, Markus; Steingräber, Peter; Swiatkowski, Gernot; Trapp, Norbert; Trunk, Dieter; Vich, Piputvat; Völker, Otto; Völzke, Ronald; Waller, Rainer; Wallner, Christian; Wendt, Fred; Wimmer, Gerhard; Zilavec, Richard.

Writing a book needs many helping hands. I would like to acknowledge my secretary, Angela Dietrich, for help and encouragement with writing the text, drawing the graphics, editing the chapters and improving the wording. Also to my students, Christian Koch, Tina Le, Christina Dörner and Natalie Alter who supported her.

I received great support from my family during the writing of this book. Thanks go to my wife Edith and our children Christian and Katrin who supported me and gave me inspiration throughout the process.

Hermann Koch

1

Introduction

This introduction explains the background reasons why the gas-insulated transmission line (GIL) will play an increasing role in the future of electric power supply and why this book is being written today.

1.1 Changing Electric Power Supply

The power transmission systems of today will see basic changes in the near future. The impact of global warming affects the structure of electric power generation. Regenerative energy sources will be used to a much greater extent than today. Onshore and offshore wind, solar thermal, photovoltaic, biomass, hydropower, geothermal and sea-based power generation using wave, tidal or under-sea current power generation are all regenerative resources to generate electric power.

The sources of regenerative electric power will enter the electric power market as large-scale generators or as distributed small-sized power installations. Large-scale electric power and generation locations are usually far away from the load centres, and need to be connected through the electric power grid [1].

To handle these new electric power resources, intelligent power flow control is needed – a so-called "smart grid". Smart energy consumption based on prices traded on the electricity market will require electric energy transmission from generator to consumer. Changing power flows caused by the availability of regenerative power sources such as wind, sun, waves, tides and all the other fluctuating regenerative energy sources will need power flow control and long-distance power transmission. Energy storage will play a much more important role if fluctuating regenerative energy is to be used efficiently. Storing regenerative energy when it is available and using the stored energy when regenerative energy generation is not available will balance the power supply. This is a requirement for the effective future use of regenerative energy.

There are several methods to store electric energy. The most common method is to pump water to a higher level when energy is available for a low price and use the pumped water to generate electric energy when the price is high. Today, such pumped storage power plants are designed to deliver peak energy for some hours a day. Large storage devices can deliver 1000 MVA for some hours, then they become empty and need water to be pumped again.

Gas-Insulated Transmission Lines (GIL), First Edition. Hermann Koch.
© 2012 John Wiley & Sons, Ltd. Published 2012 by John Wiley & Sons, Ltd.

Large storage devices can only be built in remote and sparsely populated areas, which are far away from the consumers of electricity. In Europe, for example, this would mean Norway. The available space in the Alps or Pyrenees is very limited. And lower-altitude storage does not provide the required energy capacity. Remote storage will require additional transmission lines to connect the storage with consumers.

Compressed air in large volumes in mines or underground caverns offers another possibility to store large amounts of electrical energy. Here, air will be pumped into the mine or cavern in cases when a surplus of regenerative energy is available. When no regenerative energy is available the compressed air is released through turbines to generate electricity. Today, prototype mines are under investigation in Europe and the USA, and the expectation is that this could be a relatively low-cost storage solution close to consumers where mines or caverns are available [223].

A very large storage capacity is available when the surplus of wind and solar power is used to produce hydrogen (H_2) by the electrolytic process, and store it at high pressure (about 30 MPa) in a large storage cavern in the salt layer – about 4000 to 5000 m under ground. This technology is being used today to store large quantities of natural gas. These storage capacities can be used to bridge long periods of no wind, which can occur for example in January in the North Sea for several days. The stored hydrogen can then be used with burned-in turbines to generate electricity.

Cooling or heating of residential houses or commercial cool houses is another possibility to store electric energy. Cooling or heating can be adapted to the availability of regenerative energy. When regenerative energy is available, the heating or cooling is done up to higher or down to lower temperatures. The building stores the energy using the time constant of the building. This can be part of the pricing of energy, and heating or cooling of buildings plays a role in balancing energy consumption.

Storage of regenerative energy in batteries of electric vehicles is another opportunity to balance the electric power supply. Electric vehicles stand still most of the time (80–90% of a car's lifetime is spent stationary). The batteries of electric vehicles are widely distributed and offer a large storage capacity for electric energy. Connected to the network, electric vehicles can be controlled by a smart grid to take electric energy from the network or to deliver electric energy to the network. Batteries can be used in short cycles to change from charging to delivering and can make an important contribution to a stable electric power supply based on regenerative energy sources, which are fluctuating.

Storage of surplus regenerative energy will be a key feature in the regenerative electricity market. In times when more wind or sunshine is available then the electricity market can take electric energy. This will be stored and sold later at higher prices when the electricity market requires more energy.

Small and large-scale storage facilities will be used and needed. A typical distributed small-scale storage facility is formed by the e-car batteries, as explained above. In terms of time, it covers short and mid-term sequences of minutes and hours. If the sequence of non- or low levels of regenerative energy generation reaches days or even weeks, large-scale storage facilities are needed.

Large-scale regenerating power installations installed in the near future will be far away from power consumption in metropolitan areas. Regenerative generation in offshore wind farms or desert-located solar power generation will be located in areas with low power consumption. Metropolitan areas and large cities cover only a very small part on the earth, and their density and electric power consumption is growing. Large power storage facilities will only be available far away from consumers [2, 3].

This will lead to long-distance, high-power transmission lines to connect power generation with power consumption. To balance regenerative energy sources it will be necessary to interconnect large distances, for example from the North Sea and Baltic Sea to the French or Spanish Atlantic coast; or the North East of the USA with the South East to take advantage of different wind situations in distant areas [4].

The power transmission system of today does not offer an optimized long-distance, high-power transmission network. Only relatively weak interconnections between regions are installed to cover emergency situations after the loss of large power generation (e.g., 1500 MVA) in one region when power from another region is needed.

Today, thermal power plants are located close to consumers and provide the required electric energy in a region. The future of regenerative energy generation is different, and requires an overlay network for continuous transmission of high power over longer distances, a so-called "super grid", or overlay network. The super grid will bring energy from the power generation locations to the consumers of electric power in the load centres and to the locations where electric power can be stored (e.g., hydropower pumping storage).

For other reasons, such high-power, long-distance electric power transmission has been and will be installed in China and India. Both emerging countries are building up their electric power supply by using large resources of hydropower when building dams, for example the Three Gorges in China or Tehri in India. Electric power of several gigawatts is transmitted over very long distances of 1000 km or more into the metropolitan areas of, for example, Shanghai or Mumbai.

Experience with state-of-the-art high-voltage transmission technology in China and India shows that long distances are operated over with high reliability and high efficiency – power transmission lines of ultra high voltage (UHV) for AC systems with 1000 kV AC and DC systems using ±800 kV DC. Since the year 2010 in China, ±800 kV DC is also in operation to connect large power generation stations to metropolitan areas. These long transmission lines are built as overhead lines using very high towers (typical 70–80 m) and also large bundles of conductor wires (8–10 wires in a bundle of about 500 mm to 800 mm diameter). Such large overhead lines may not be possible everywhere, and underground solutions for at least sections of the transmission line may be needed. The need for high-power transmission and the possibility of combining overhead transmission lines with underground solutions make the GIL a good candidate for an overall solution, or at least part of the solution [217].

Today, large-scale wind farms onland and offshore are in planning or under construction. Large-scale photovoltaic generation or solar thermal generation is in planning and the first pilot projects are proving feasible. While wind is mostly best at sea and sun shines best in deserts, long-distance transmission is needed to compensate for fluctuation and to bring the regenerative energy to the consumers.

When overhead lines cannot be built, then the GIL offers an alternative solution to going underground with the same amount of energy as the overhead line can transmit (e.g., at 400 kV the transmitted power is 2000 MVA).

More than 40 years' experience of first-generation gas-insulated technology is an excellent basis for using GIL as a high-power transmission system underground. The gas-insulated technology was introduced to substations in the late 1960s and is widely used today with high reliability. The introduction of second-generation GIL in 2001 using N_2/SF_6 gas mixtures and pipeline laying methods to reduce costs makes the GIL a long-distance, high-power underground transmission system with high reliability and availability for the future [168].

1.2   Advantages of GIL

GIL offers several advantages for high capacity power transmission, as listed below [9, 10].

Low Transmission Losses   Resistive losses are low because of the large cross-section of the conductor and enclosure pipes. Typical GIL resistances are 6–8 mΩ/km depending on the outer diameter (500 mm or 600 mm) and the wall thickness of the enclosure and conductor pipe (6 mm to 15 mm). The transmission losses are related to the square of the transmitted current as $P_v = I^2 \cdot R$ (I = current, R = resistance). When the current rating is high – as it is for GIL (e.g., 3000 A) – then the effect of low transmission losses is high. The losses through the insulating gas are negligibly small.

Low Capacitive Load   Electric phase-angle compensation is only needed at very long lengths, because the capacity of the GIL is low, typically 55 μF/km. Therefore, no or low compensation coils are needed under most network conditions for transmission lengths of about 100 km. This also reduces the thermal operation losses.

Power Rating Like an Overhead Line   The GIL is the ideal alternative or supplement to overhead lines. The high power transmission capability of the GIL (up to 3000 MVA per system at 550 kV rated voltage) allows it to go directly underground in series with an overhead line without power reduction. The GIL also allows the use of protection and control systems in the same way as with overhead lines. No differential protection is needed for failure location when a GIL is combined with overhead lines. The GIL has a low capacitance and, therefore, the inrush current is low.

High Level of Personnel Safety   The outer enclosure pipe is solid grounded and no access to high-voltage parts is possible (gas-tight enclosure). Personnel safety is also guaranteed in case the GIL has to carry a short-circuit current (50, 63 or 80 kA up to 1 or 3 s). Even in case of internal failure and an arc between the enclosure and conductor pipes, tests have shown that no external impact occurs on the surroundings.

High Reliability   The only purpose of the GIL is electric power transmission. No internal switching or breaking capability is needed. Based on this, the GIL can be seen as a passive high-voltage gas-insulated system with no active moving parts (e.g., switches). Today, more than 300 km of single-phase lengths has been in operation world-wide for more than 35 years. So far, no major failure (arc fault in the system) has been reported. This makes the GIL the most reliable power transmission system known.

No Electric Ageing   Gas insulations do not age. The best example is an overhead line with ambient air as insulating gas. The electric field strength of the insulators and the maximum temperature of the GIL are too low to start the process of electrical or thermal ageing. This has been proven using long-term measurements in independent laboratories and also by extensive experience with the equipment in the network. The first GIL installations have been in operation since 1974, and the results are reported by the CIGRE [71, 224].

Operation Like an Overhead Line   Overhead lines in the transmission network are operated with the so-called autoreclosure function. This means that in case of a ground fault detected on the line, the circuit breaker will automatically break the lines, wait some seconds (depending on the network condition) and then switch on again. In most cases the reason for the fault current detection will be gone and the transmission line will go back to normal operation (for example, if a tree branch gets too close to an overhead line, the branch will be burned away or if a lightning strike causes the fault current, that will also be gone after some seconds).

Electromagnetic Fields   To protect the public and the operational personnel international regulations require electromagnetic field limitations. These values vary across regions and countries depending on laws and regional regulations. A trend can be seen worldwide that limiting values are getting lower and the restrictions harder. In densely populated areas and cities these electromagnetic field requirements are defining the allowed design of transmission lines.

The GIL is operated as a solid grounded installation and the inductive loop is closed through the ground connection. The coupling factor is about 95%. This means that the superposition of the two reverse currents reduces the outside magnetic field by 95%, and only 5% of the magnetic field of the conductor current is effective outside the GIL.

Because of the induction law, the current in the conductor will induce a current in the enclosure of the same size and with 180° phase shift. The superposition of both electromagnetic fields is close to zero. In case of limitation of the magnetic field in the surroundings, this solid grounded GIL can fulfil even very low magnetic field requirements. With a current rating of 3000 A, within a few metres' distance a magnetic field strength of 1 µT can be reached (as required in some countries).

The advantage of a low magnetic field is important when residential areas are close to the transmission line – for airports with their sensitive instruments, hospitals with their sensitive imagining systems, or all kinds of sensitive electronic equipment in private or business use.

In Italy, electromagnetic field requirements for new installations go down to magnetic flux values of only 0.2 µT. When residential areas are involved, the GIL can reach such low values over a distance of a few metres.

No Thermal Ageing   The GIL is designed for maximum operational temperatures given by the surrounding conditions – maximum 60 or 70°C touching temperature in a tunnel, or 40 or 50°C when directly buried. The different temperature values depend on individual countries and their applied standards and regulations.

In all cases the maximum allowed temperature of the conductor of 100 to 120°C is not reached by far. Therefore, no practical ageing of the system can be expected under these operating conditions.

2

History

In this chapter the historical background of electric power supply and the development stages of the gas-insulated transmission line will be explained.

2.1 Transmission Network Development

2.1.1 General

When electrification started in the mid-1800s, electric power generation was installed close to the consumers. At voltages of about 100 V DC, generators delivered electric power to the consumers – for electric lights and electric drives. The generators were driven by hydropower at rivers, and the consumers were households, farms, offices and small industry (e.g., sawmills).

When cities started using electric street lights and manufacturers started using electric machines for production, electrification grew quickly. Electric generators used running water at rivers or steam engines in power houses transmitted the electricity using DC. Point-to-point connection was the normal case, mainly because it was difficult to switch DC currents. During these early periods of electrification, discussion was rife over the relative advantages of DC vs AC. Arguments led to two main positions:

- Tesla position – favoured AC, because it can be transferred to other voltage levels and is easy to switch.
- Edison position – favoured DC, because of its low transmission losses and higher transmission efficiency.

Today we know that AC won the battle in this first development stage of the electrical network, and the main reason was the transformer for higher voltage levels and the availability of AC switching devices. When AC was transformed to higher voltage levels of some kilovolts, and switches and circuit breakers managed to operate reliably, the extent of electricity use increased. Larger power generation units were possible to serve more consumers in a switched distribution network of high AC voltages.

Gas-Insulated Transmission Lines (GIL), First Edition. Hermann Koch.
© 2012 John Wiley & Sons, Ltd. Published 2012 by John Wiley & Sons, Ltd.

With AC technology developing new materials, higher voltage levels were introduced and could be managed. With improvements in reliability and service life of electric light bulbs and electric motors, the number of installations and with that the electric power consumption increased. The voltage levels went to higher values and reached some kilovolts in the late 1800s. At this time, the first central power stations using steam engines to operate AC generators were installed in cities like New York, London, Paris and Berlin.

With the technical development of insulating materials, the production of insulators and the better understanding of high-voltage electric fields, higher operating voltages were introduced and with the availability of switches and circuit breakers the point-to-point electric power transmission system developed into an electrical network with high-voltage power transmission and medium-voltage distribution. Transmission voltages went into the 100 kV range and small local power producers cooperated and connected their electrical systems to develop city-sized power producers and distributors. They formed the first distribution companies and became known as public utilities.

Based on the growth of these public utilities the units of electric power generation increased in size, and in cooperation with other utilities larger areas were served with electric power supply. Larger power plants of 100 MW to 500 MW, and later 1000 MW, were developed and installed. With this larger power generation the power transmission voltages also increased, with new transmission voltages of 220 kV typically in Europe or 242 kV typically in North America. These transmission voltage levels came into use at the beginning of the 1920s. The next step increase in transmission voltages came in the 1960s, with 380 kV in Europe and 345 kV/550 kV systems in North America and Japan. Over the following years, transmission voltage levels of 345 kV up to 550 kV were installed all over the world. The highest voltage levels for transmission networks occurred in North and South America, South Africa and Russia – reaching 735 kV, 765 kV and 1000 kV. Today, activities on UHV transmission lines can be found in China (1100 kV) and India (1200 kV). Plans are also underway in North America and Europe to connect distant regenerative energy sources to the load centres by UHV lines [12].

2.1.2 Power Transmission Levels

The electric power system of today can be assigned to five levels: local, regional, national, international and intercontinental, as shown in Table 2.1 [13, 14].

Local Level   The local level (level 1) of the electric power system has close reach to the consumers, with a typical distance of 1–5 km. The local level represents the origin of electric power supply, when 100 V DC was used in the first place. Then, the development was from DC to AC with typical house connection phase-to-ground voltages of 110 V AC in, for example, the USA and Japan and 240 V AC in, for example, Europe and some Asian and South American countries. Three-phase electric systems of 400 V AC are used today. These voltage levels are in wide use and cannot easily be changed because of the high cost of replacing all the installed equipment.

Regional Level   The regional level (level 2) can be identified as covering a region, typically a city and the surrounding area. The reach of such a region is typically up to 100 km. The

Table 2.1 Power transmission levels

Level	Type	Example	Typical distances	Typical voltage levels	Typical current
1	local	small power producer city, e.g. Berlin, London, New York, Paris	1–5 km	up to 500 V	500 A
2	regional	utility city, e.g. Metropolitian areas of Berlin, London, New York or Paris; region, e.g. Bavaria	100 km	1–52 kV	2000 A
3	national	utility country, e.g. France, Italy; larger region, e.g. North East USA	1000 km	up to 400 kV* and 550 kV**	4000 A
4	international/ interregional	interconnected network, e.g. UCTE, NORDEL, MAGHREB, Itaipu, Cabora Bassa	2000 km	400 kV* 550 kV** 800 kV** 1100 kV***	5000 A
5	intercontinental	intercontinental networks, e.g. Europe/Africa, Europe/Asia	3000 km	1200 kV****	6000 A (expected)

* Europe, Middle East, India, Africa.
** North/South America, Asia.
*** China, Japan.
**** India.

voltage levels are below 100 kV and large power generation units are installed inside this region. Power producers in a region may be private or public entities owned by the city or regional organization. These larger companies, compared with the local power producers, own larger power generation units and operate higher voltage level transmission and distribution lines to serve more customers in a larger area. At the regional level several voltage levels have been introduced: 1 kV, 3 kV, 4.5 kV, 12 kV, 15 kV, 20 kV, 33 kV, 42 kV, 52 kV, and some more voltage levels between these values can be found around the world. These many steps in voltage levels represent the technical development of insulating materials. Today, the international standardization defines distribution voltages as 1–52 kV.

Many private companies and public-owned utilities were formed when the regional power transmission and distribution structure was built, and these companies are still in operation today – serving individual cities or regions. The typical reach of such a regional electric power supplier is about 100 km. Today, the deregulation of the power market has the goal of unbundling the monopolistic structures of the past, changing ownership and creating new structures of regional power supply under competition. The goal is for electric power users to have a choice of different power deliverers, with regulations and laws set in place. This will bring new players into the electricity market, and more companies will offer and trade with electricity. Trading with electricity was never in the mind of engineers when the power transmission and distribution network of today was designed. This means a basic change in the design and use of the power network.

National Level   The national level (level 3) of electric power supply can often be identified with national borders of countries (e.g., France) or parts of countries (e.g., California in the USA). In some countries, the electric power supply was seen as a national task and public or government-owned power companies have been formed with governmental regulations. In Europe, such national level power suppliers have been established with, for example, EDF in France or ENEL in Italy. In other countries, with a more federal structure, the development of electric utilities was via private entities. These private entities grew to the same size in terms of generated megawatts or number of served customers as the public utilities. In Europe, such power supply companies can be found, for example, in Germany (e.g., RWE, PreussenElektra, Bayernwerk). In North America, regional utilities can reach the same area as whole countries in Europe (e.g., New York Edison in the Northeast, Bonville Power Association [BPA] in the Northwest, or Tennessee Valley Authority [TVA] in the Southeast).

The typical reach of a national power supply level has distances up to 500 to 1000 km and in consequence, voltage levels reached values of 420 kV in Europe, the Middle East, Africa and parts of South America and Asia and 550 kV in North America, Japan, parts of South America, Asia and Africa.

International Level   The international or interregional level (level 4) can be identified with an area of several countries or larger regions. The level 4 network is used to stabilize the power transmission network in total and to offer network stability. Historically, the development of level 4 was meant to connect regions with high-voltage lines for emergency situations when power generation was lost in one country or region. In case of power losses because of interruptions in generation or interruption of transmission lines, the connection to the neighbouring region supports the network and keeps it stable. The UCTE in Europe developed the rule that the power of two 1500 MW nuclear power units can be compensated by international or interregional power reserves of 3000 MW. At each point in the network the concentration of electric power is limited to this value. The typical reach of this international or interregional level is about 3000 km, considering networks in Europe and North America.

In Europe, the northern NORDEL, the central UCTE and the eastern Baltic networks are synchronized in connection. In the USA, independent system operators (ISOs) have been established. These ISOs are operated with fixed frequencies and have interregional connections for relatively low power capacities.

Intercontinental Level   The intercontinental level (level 5) does not really exist today but is on the way to being developed following the need for long-distance power supply when regenerative energy sources are in use world-wide. This layer needs even higher voltages to cover the long distances of up to 3000 km of electric reach for such intercontinental power transmission. These very long distances require the advantages of high-voltage DC power transmission, which offers low power transmission losses. As part of this future intercontinental power transmission level, the first installations are already in place in different parts of the world. The intercontinental level is certainly 30 years or more into the future, but current projects in China and India show that it is technically and economically feasible.

2.1.3 Long-Distance Power Transmission

In Africa, Cahora Bassa was installed in 1977–1979 to connect the hydropower resources of the Sambesi River to industrialized South Africa with a length of 1420 km. The power transmitted is at 1920 MW and the operational voltage is ±533 kV DC (see Table 2.2).

Table 2.2 Main data of Cahora Bassa, South African Development Zone [8]

Commissioning	1979, upgraded 2008
Power rating	1920 MW
DC transmission voltage	±533 kV
Length of overhead line	1420 km

   In Figure 2.1 a large hydro dam and in Figure 2.2 a small hydro dam is shown. Large dams are often located in remote areas and need connection to the load centres by high-voltage AC or DC transmission lines. Small dams are usually located closer to load centres and are connected by lower high-voltage levels.

   In Itaipu in South America, hydropower generated in the Argentina/Uruguay border area is delivered to the region of Sao Paulo in Brazil over 800 km away. The hydropower station at

Figure 2.1 Example: Large hydro dam. Reproduced by permission of © Siemens AG

Figure 2.2   Example: Small hydro dam. Reproduced by permission of © Siemens AG

Itaipu went into operation in 1987, and with two high-voltage DC systems a total of 6300 MW is transmitted to the electric load centre in and around Sao Paulo (Technical data of the Itaipu project is shown in Table 2.3. Figure 2.3 shows the 735 kV AC gas-insulated switchgear [GIS] inside the dam).

One of the early projects in the USA was the Pacific Inter-Tie, which transmits a total of 3100 MW from Oregon's water dams to the metropolitan area of Los Angeles. The distance is about 1360 km, see Table 2.4.

Table 2.3   Main data of Itaipu, Brazil/Argentina [16]

Commissioning	1984–1987
Power rating	6300 MW
DC transmission voltage	±600 kV
Length of overhead line	Line 1: 685 km
	Line 2: 805 km

Figure 2.3   High-voltage gas-insulated switchgear in a hydropower station. Reproduced by permission of © ABB Switzerland Ltd

In several steps from 1970, 1985, 1989 and 2004, high-voltage DC transmission was installed and upgraded each time to higher power ratings from 1440 to 1600 to 2000 and 3100 MW as of today; the DC voltage rose to ±500 kV.

In Figure 2.4 an ±800 kV DC overhead line is shown. The tower height is about 80 m and the span width between towers is about 500 m.

In China, high-voltage DC long-distance power transmission lines have been built at Central China's Gezhouba on the Yangtse River to transmit 1200 MW over 1040 km to the Shanghai region.

In Southern China, several DC lines are in operation to bring hydro-generated energy from the mountainous regions around the cities of Kunming and Guiyang over 1000–1500 km into the industrial centres of Shenzhen and Guangzhou in Southeast China. A total power of more than 12000 MW is transmitted over several overhead lines. A new step in the technology of high-voltage DC power transmission started in 2010, with a higher transmission voltage of ±800 kV DC [18].

Table 2.4   Main data of the Pacific Inter-Tie [17]

Commissioning	1970, 1985, 1989 and 2004
Power rating	1440 MW > 1600 MW > 2000–3100 MW
DC transmission voltage	±400 kV > ±500 kV
Length of overhead line	1360 km

Figure 2.4   ±800 kV DC overhead line. Reproduced by permission of © Siemens AG

In Figures 2.5 and 2.6, outside and inside views of the Gui-Guang I converter station are shown. In Figure 2.7, an overview graphic gives an impression of the length of the Gui-Guang II DC transmission line [196].

Technical data for Gui-Guang I is given in Table 2.5, with a 3000 MW power rating at ±500 kV DC.

On the AC voltage side, for long transmission lines a voltage level of 1100 kV was introduced in China on a 600 km long transmission line in the year 2009.

In India, some long-distance lines have been installed to get electric energy into the large metropolitan areas of Mumbai, Delhi or Kolkata. More high-voltage DC lines are under consideration, and investigations into new 1200 kV AC systems are underway. Figure 2.8 shows the long-distance DC transmission line called the East South Interconnector in India. Technical data is given in Table 2.6.

In China and India, some more long-distance and high-power transmission lines are planned. The main reason is to bring energy from remote areas of regenerative resources (e.g., hydropower dams) to the metropolitan and industrial centres far away.

In Europe, the future of large-scale regenerative power generation will also need long-distance transmission solutions. The offshore wind farms in northern Europe and solar power generation in southern Europe and North Africa are also far away from the consumers of electricity and will need new long-distance high-power transmission lines.

In North America, new solar and wind generation will be built in desert countries like Arizona or wind countries like Montana, or even offshore locations for large wind farms. All

Figure 2.5 Long-distance DC transmission – Gui-Guang I, China. Reproduced by permission of ©
Siemens AG

Figure 2.6 Converter hanging from the roof, ±500 kV DC. Reproduced by permission of © Siemens
AG

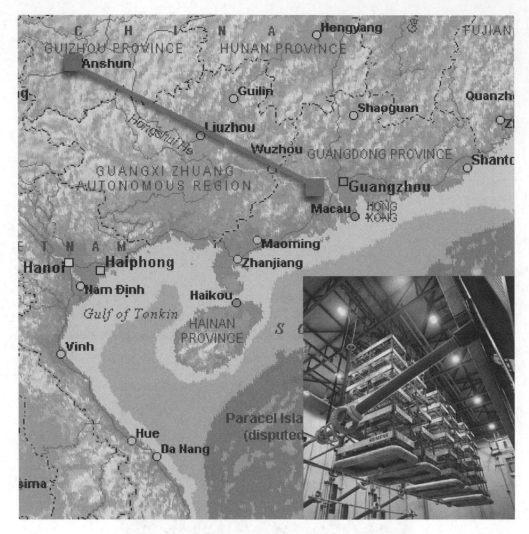

Figure 2.7 Long-distance DC transmission – Gui-Guang II, China, 2007. Overhead lines and filters next to converter building. Reproduced by permission of © Siemens AG

Table 2.5 Technical data for Gui-Guang I, China [19]

Rating	3000 MW
Voltage	±500 kV
Line length	940 km
Contract	1 November 2001
Project completed 6 months ahead of schedule by September 2004	
Thyristor: 5″ LTT with integrated overvoltage protection	

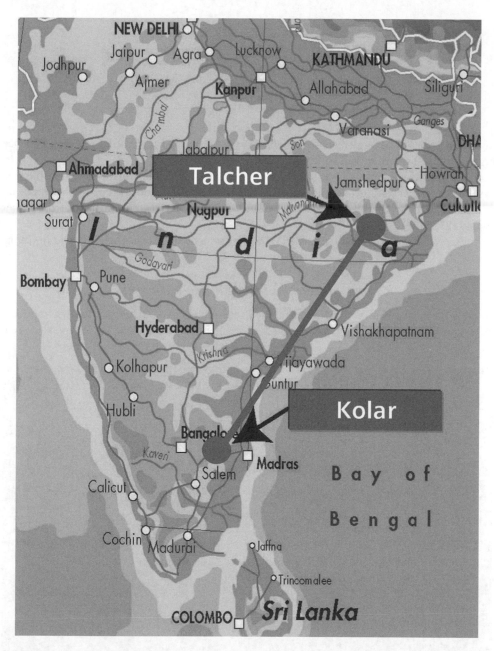

Figure 2.8 Long-distance DC transmission East South Interconnector, India. Reproduced by permission of © Siemens AG

Table 2.6   East South Interconnector, India [20]

Rated power	2000 MW, bipolar
DC transmission voltage	±500 kV
Rated current	2000 A
Distance of transmission	approx. 1450 km

of these regenerative energy resources are far away from the load centres and new long-distance high-power transmission will be needed.

These new installations can be seen as part of a future intercontinental power transmission system. The projects follow the same goal, of bringing renewable electric energy (hydropower, wind, sun) to the distant load centres and building up a regenerative, sustainable electric power supply [21, 22].

In Figure 2.9, the world's first ±800 kV DC converter station is shown – installed in China.

2.1.4   Current Ratings of Electric Transmission Networks

Besides the increase in voltage levels, the current ratings are also increasing. At the local level 1, typical current ratings are below 500 A and at the regional level 2, typical current

Figure 2.9   ±800 kV DC converter station in China. Reproduced by permission of © Siemens AG

ratings of 2000 A can be found. The national level 3 shows typical current ratings of 3000–4000 A, with a tendency in some regions towards 5000 A. The international or interregional level 4 shows current ratings of 5000–6000 A and, finally, the intercontinental level 5 might have 6000 A and maybe 7000 A (when the current projects in China are taken as reference and with development planning in mind).

The increase of currents is, besides the increase of voltage, a second way to increase the power transmission. Two main problems are related to increased currents: heat and mechanical forces.

Heat is related to the transmission losses in the conductor, which create heat and higher temperature. Maximum temperatures of contact materials are usually around 100°C. When higher temperatures are required, the price of materials goes up quickly.

Heat is produced at rated currents of constant flow, which is the normal operation status of the transmission system. These currents are the design values for diameters and wall thickness, and are dimensioning part of the transmission system. They are also a measure of the cost: the more material, the more cost.

The second heat dimensioning comes from the short-circuit ratings, which can have currents of 63 kA, 80 kA or even 100 kA for 1 s or 3 s, depending on the network configuration. These extremely high current values increase the heat – mainly in any contact system – and may increase the temperature above limit values.

Besides the heat, the electromagnetic forces between the conductors need to be withstood by the design, which may result in costly design measures (e.g., increasing the number of fixing insulators).

The current rating of electric transmission networks limits the increase in power transmission. The costs of design might increase strongly, and the technical solution might become uneconomical. The current rating is linked to the dimensioning, which is why with each step to a higher voltage rating the current rating also takes a step up.

2.1.5 Conclusion of Transmission Network Development

The more than 150-year history of electric power supply, from the first electric street lights in Boston, New York, London, Paris and Berlin in the middle of the 19th century, has seen tremendous changes and steps in development. Today, electricity is the most important energy source for modern life. How can we live without electric power supply? Life would come to a halt. For the future, the importance of electricity is increasing.

Renewable energy sources will show a great increase in the coming decades. The reduction of CO_2 with electrical energy generation will support regenerative energy sources. The change – increasing wind and solar energy sources and decreasing coal, gas and nuclear – in the long run will have a great impact on the power supply network [23].

From an historical point of view, it has been shown here that the existing structure of electric power supply will not meet the requirements of the future. Until today, the electric power network has been designed for close-to-consumer located power plants. This will change in a regenerative network, to far-from-consumer located energy sources.

This change will require, in consequence, long distances of power transmission at higher voltage levels and higher current ratings. An overlay transmission network at 800 kV, 1000 kV or even 1200 kV will be required. Transmission network nodes of high power (e.g., at nuclear power plants) will connect points of this overlay network. A new transmission network level will be introduced, to act as an intercontinental transmission network. It might be called a "super grid" and is based on ultra high voltage (UHV). An information source for these activities can be found at the website of "The Friends of the Supergrid" [197, 225].

This network change needs new solutions. Besides the existing technologies of overhead lines and solid insulated cables, the GIL offers an additional solution for high-power transmission. The GIL offers an opportunity to transmit large amounts of electrical energy over long distances directly buried, underground or installed in a tunnel.

The voltage range for GIL covers high voltages from 100 kV up to 800 kV. Most applications of the GIL are at transmission network voltages of 420 kV and 550 kV. The upper range of 800 kV finds only a few applications in China today. Today, the long-distance, high-voltage transmission lines are related to large hydropower plants. For example, in North America from the great lakes in the Canadian north to the cities on the northeast coast of the USA. In South America from Itaipu to the cities on the coast. In Russia from the rivers in Siberia to the Moscow region. In China from the mountains in the west to the cities in the east, or in India from the mountains in the north to the cities in the south and central area. In some other cases large fossil power plants are the reason for long-distance transmission. In the USA from the southwest and southern coal regions to the cities in the north along the Pacific coast. In China from the northeast coal region to the metropolitan areas of Beijing and Shanghai. In India from the central subcontinent to the northern metropolis of Delhi and the western metropolis of Mumbai.

In Europe, long-distance transmission lines are under development and planning. With the driver of renewable energy generation of large wind farms in the north and large solar power generation in the south of Europe and North Africa, the intercontinental level 5 is under development.

Most of these long-distance transmission lines will be built as overhead lines because this is the lowest-cost solution. But in some densely populated areas or areas under environmental protection, underground solutions are necessary. The GIL offers an alternative solution for these high voltages and current ratings [25, 26].

In the end, the intercontinental electric power transmission network might be the solution for a sustainable power supply based on regenerative power generation world-wide. See the map in Figure 2.10.

2.2 Historical Development of GIL

2.2.1 GIL 1st Generation

The first-generation GIL is filled with 100% SF_6, and might be welded or flanged, but without elastic bending of the aluminium pipes.

2.2.1.1 Introduction

What was the reason to develop a GIL?

The basic reason for the development of the GIL is based on an artificial gas: sulphur hexafluoride (SF_6), with very good electrical insulation properties. Found and designed in the early 1920s, SF_6 is an artificial gas which is non-toxic, inert, non-flammable, non-corrosive

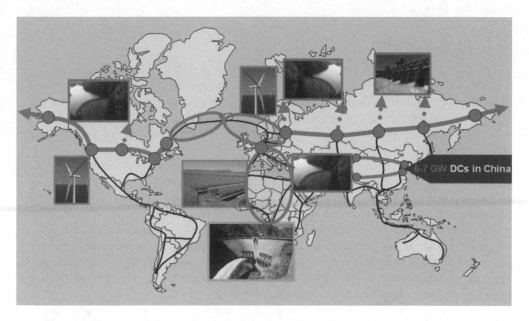

Figure 2.10   Map of ultra-high voltage lines. Reproduced by permission of © Siemens AG

and long-time stable. Its electrical properties allowed economical technical systems to be designed for high voltages and high current ratings. At a time when the existing oil technology had reached its technical limitation, SF_6 insulation opened up a new development step into an electrical future [27–31].

In the 1960s, the first experimental set-ups were investigated using SF_6 in closed compartments under high-voltage conditions with DC and AC voltages. The physical nature of a pipe-type conductor of aluminium, and a pipe-type enclosure also of aluminium, held by insulators and filled with SF_6, was seen as the best solution for high electric power transmission. For long distances, DC has lower transmission losses than AC. Therefore, DC has been investigated to design a GIL in the first place.

At this time, DC caused many problems in dielectric stability of the insulating system. Some physical phenomena – like space charging of insulators or surface charges on insulator surfaces – were new and not fully understood. In consequence, the GIL product development went to AC systems. The AC GIL product development was more successful and delivered good results. On this basis, gas-insulated technology was introduced in the substations. The first applications were sections of bus bars without any switch. Switchgear soon followed, because of the very good arc-distinguishing properties of SF_6.

The very good insulating capability of SF_6 reduces the dimensions of high-voltage systems. The excellent arc-interruption capability increased the switching capability of circuit breakers or switches like no other insulating gas today known. This led to investigation for SF_6 circuit breakers, which were first applied in the high-voltage network at the beginning of the 1960s with the first complete SF_6 switchgear including circuit breaker, ground switch, disconnector and bus bar in 1968. This SF_6 technology developed in the market to a global, very successful, high-voltage switchgear product of high switching capability and reliability. Today, thousands of GIS bays are installed world-wide, serving all voltage ranges of the power network. With increasing power ratings and limited space available, GIS is expected to increase the market share in the future.

The development of GIL went in parallel with GIS. The successful use of SF_6 as an insulating medium, and the high capacity of electric power transmission, the GIL was a new choice in the network. The GIL as a transmission system is seen as an alternative for overhead lines when these cannot be built because of non public acceptance or environmental restrictions. The GIL is a solution in sections or for the total length of the power transmission network when the power rating is high, e.g. 2000 MVA.

In parallel with the GIS development, different GIL technologies have been developed. GIL jointed with O-ring sealings and flanges, or jointed by weldings. With the different designs, different names were created: gas-insulated bus duct, compressed gas-insulated bus duct, bus bar, gas-insulated transmission line (GITL) or gas-insulated line (GIL). These names are in world-wide use today. In 1998 the international standardization organization IEC introduced the name "Gas-Insulated Transmission Line" with abbreviation GIL as the preferred term for use world-wide. The preferred name GIL is for any gas-insulated system, using atmospheric pressure or overpressure inside, which has no switching or breaking function and is longer than 500 m. This is the actual definition of IEC for GIL.

The GIL used inside substations and in conjunction with GIS is jointed by flanges and the sealing for gas-tightness is made by O-rings, the same technology as in GIS. When used as power transmission lines in tunnels, above ground or directly buried the joints are welded. There are different types of enclosure and conductor pipes used. In AC they are made of aluminium or aluminium alloys. The pipe may be made from aluminium plates, sheet material or raw material for the extruding process.

The plates are formed by a mechanical triple-roll mill and then the pipe is closed by a longitudinal weld. The sheet material uses coils for spiral welding the pipe in an endless production process. The raw material for the extruded pipe is a block of aluminium alloy which is then formed into a pipe using an extruder. These extruded pipes are seamless. Two welding technologies are used: hand welding and automated orbital-welding machines.

The typical application voltage is greater than 100 kV up to 800 kV with the majority of installations at the 420 kV and 550 kV voltage levels in transmission networks. The first GIL installation world-wide was in the year 1974, going into operation in 1975 at the Cavern Hydropower Plant of the "Schluchseewerke" in the Black Forest in Germany at a voltage level of 400 kV. This type is called: "The first generation of GIL" [49, 74].

2.2.1.2 First 400 kV Project: Schluchseewerk

What was the reason for this first-generation GIL project?

The pumping storage power plant at Schluchseewerke plays an important role in the German 400 kV transmission network in stabilizing the voltage and frequency. In sequences of hours the plant is used to generate up to 1000 MW (generating mode) by letting water run from the reservoir on top of the mountain through turbines in the cavern of the mountain and deliver energy to the network with two transmission systems. Or, in case of an energy surplus, to get 700 MW (pumping mode) from the network and to pump water up the mountain into the reservoir [31].

Table 2.7   Technical data of the Schluchsee project, Germany

Type	Value
Nominal voltage	380 kV
Maximum voltage	420 kV
Nominal current	2000 A
Lightning impulse voltage	1640 kV
Switching impulse voltage	1200 kV
Power frequency voltage	750 kV
Rated short-time current	135 kA
Rated gas pressure	7 bar abs.
Insulating gas mixture	100% SF_6

A failure in an oil cable destroyed not only the cable but also the tunnel. During the replacement of the destroyed tunnel the owner was looking for a non-burnable transmission system and came to the GIL as the solution. GIL connects the high-voltage transformers at 420 kV with the overhead line through a 700 m long tunnel.

The technical data for the Schluchsee project is shown in Table 2.7.

In Figure 2.11 the principle set-up of the Schluchsee hydropower plant is shown.

In each cavern a 420 kV high-voltage machine transformer (1) is located with a power capacity of 600 MVA. This machine transformer connects the power generator with the GIL. Close to the cavern for each GIL system, gas-insulated surge arresters (2) are located to limit the overvoltage on the generator side and at the overhead line towers additional surge arresters

Figure 2.11   Principle set-up of the Schluchsee GIL. Reproduced by permission of © Siemens AG

Figure 2.12   View into the tunnel. Reproduced by permission of © Siemens AG

(5) are located to prevent the GIL from overvoltages coming from the overhead transmission (e.g., because of lightning strokes).

Between the vertical sections of the two GIL systems (1a) when leaving the cavern to the GIL systems (1b) in the tunnel a disconnecting switch is located to be able to separate the overhead line (6) side from the generator sides at the transformers (1).

Figure 2.12 shows a view into the tunnel at the slope with two GIL systems fixed to the side walls of the tunnel. The tunnel is built into rock and has a width of 2.8 m and a height of 3.5 m. The GIL pipes are fixed to steel structures on the tunnel sides and laid on rolls for movements related to the thermal expansion. The 7 m long GIL sections are assembled on-site and the joints are made by orbital welding. To guarantee the small tolerances in diameters and roundness, which are required to meet the high-voltage tolerances and clearances, the hand-welding process has to follow precise welding sequences [33].

The 420 kV GIL was, in 1974, the first gas-insulated installation world-wide and, therefore, the test voltages have been chosen high, just to be on the safe side. The value for lightning impulse tests is given at 1640 kV, today after 35 years of experience the international standard value is 1425 kV. The same can be found for the switching impulse voltage, with 1200 kV versus the 1050 kV level of today and for the power frequency test voltage, with 750 kV for the Schluchsee project against today's level of 630 kV (see Table 2.7).

With these precautions the new GIL technology of the first generation went into service in 1974 at the Schluchsee Pumping Storage Power Plant in the Black Forest in Germany, and is still in operation without any major failure.

A second reason for higher test voltage levels is the high lightning probability in the Black Forest in South Germany. Lightning strikes cause transient overvoltages, which travel along the overhead line and enter the GIL. These additional voltage stresses have been covered by choosing higher test voltage levels.

The design engineers who developed the first-generation GIL knew that the world was looking for the success of this project; it should not fail. As a consequence, they chose higher test voltages which could later be reduced, when more experience with the new technology at the 420 kV voltage level was available.

Today, we can say that the GIL at Schluchsee has been in continuous operation for more than 35 years without problems and the gas-insulated technology is established as very reliable high-voltage equipment.

What can we learn?

Once gas-insulated systems are installed and in operation they stay reliable for a very long time without any sign of aging.

- Pioneering work was necessary to design the first-generation, single-phase, SF_6-insulated, gas-insulated equipment, called GIL.
- The first-generation GIL found its main field of application in substations or power plants. Typical applications are the connection of gas-insulated switchgear (GIS) with overhead lines or transformers.
- In hydropower plants, to connect the power generation unit on the high-voltage side through a tunnel to the overhead lines of the network.
- For long-distance application in the transmission network the absolute cost of the first-generation GIL was too high. This led to the development of the second-generation GIL [46, 49, 74].

2.2.1.3   Experiences with 1st Generation GIL

The first world-wide application of a 420 kV GIL was built as the connection of a hydropower pumping storage plant in Schluchsee, Germany, with an overhead line outside the mountain for the connection of the transmission network. The GIL was constructed in the years 1974 and 1975 and now has a successful history of 35 years. It has a system length of 700 m, over two systems. It can transmit 2000 A. This installation runs through a tunnel in the mountain and is used as a peak load hydropower generation station. In low load times at night the water is pumped up the mountain to the water storage and at high load times during the day the water generates electricity to cover the load peak. So, most of the time the GIL is running at rated currents for pumping or for power generation day by day. The GIL has operated at a very high reliability since its installation until today and has run without interruption for more than 35 years. For the planned 25-year revision which took place in the year 2000, it was decided by the user and the manufacturer not to carry out any maintenance work on the GIL. The revision date was extended by another 10 years, with a next check in 2010. Practically no ageing effects have been detected and no indication of ageing was given by the operation protocols. This experience is also in line with the results of the CIGRE study committee SC15. In their report it was stated that for gas-insulated systems practically no ageing can be assumed [224].

Gas-Tightness   The operational experiences with the GIL at Schluchsee have shown that there is no gas leakage detected when the pipe joints are welded. The gas-tightness of the GIL did not require any gas refill for the 35 years of operation. Therefore, the on-site welded GIL can be seen as gas-tight.

Overvoltage and Surge Arresters   Five years after the commissioning, an extreme thunderstorm and a lightning strike directly into the overhead tower next to the GIL bushings caused a flashover in the GIL and an interruption of power transmission. The main reason for this incident was the suboptimal placement of the overvoltage surge arresters in the transfer cavern in the mountain because of the large size of the zinc oxide surge arresters. With this experience and the availability of much smaller MOX surge arresters, an optimal placement at the best tower of the overhead line could be chosen and no other failure occurred in the following 30 years of operation.

What has happened?

In June 1980 a very strong lightning strike into the last tower of the connected overhead line caused a failure current larger than 100 kA into the L2 conductor. The transient overvoltage surge entered the GIL and was not recognized by the zinc oxide surge arresters in the surge arrester cavern in the mountain. The transient overvoltage surge reached the transformer bushing at the end of the GIL and, because of the reflexion factor, the overvoltage doubled. The transient double overvoltage surge now travelled back on the unprotected section of the GIL between the transformer bushing and the zinc oxide surge arrester in the arrester cavern. This caused a flashover in the GIL and the protection switch-off of the power.

The GIL needed repair by exchanging the damaged parts and went back into service again after the high-voltage tests were passed. To prevent the GIL from such an overvoltage in the future, new MOX surge arresters have been placed before the bushing of the GIL. Since then, no further incident has been reported and this solution is today used in all cases where the GIL is directly connected to the overhead lines. The conclusion is that surge arresters are needed when the GIL is connected to an overhead line at the transition point from the overhead line to the GIL.

2.2.2   GIL 2nd Generation

The second-generation GIL is filled with a gas mixture of, for example, 80% N_2 and 20% SF_6, it may be welded or flanged, and elastic bending down to a radius of 400 m is used [206].

2.2.2.1   Prototypes of the 2nd Generation

What was the reason for the development of a second-generation GIL?

Why did a second-generation GIL become necessary?

There are two main reasons:

> Reason 1. As explained before, the future transmission network will need more underground solutions. New transmission lines for future regenerative energy generation and new long transmission lines will need the GIL as a technical alternative solution.

> Reason 2. The cost of the first-generation GIL needs to be cut by 50% to make long-distance projects economical. This has been achieved by a new laying process and technology adopted from pipeline laying.

Table 2.8   Key technical data for GIL, from EDF feasibility study (1994) [36, 37]

Power per three-phase system	3000–4000 MW
Nominal voltage	400 kV
Maximum voltage	420 kV
Current rating	4300–5700 A
System length	up to 100 km
Isolating gas mixture	$N_2 > 80\%$ and $SF_6 < 20\%$
Type of laying	directly buried

Both reasons were the basis for development programmes in 1994 to find technical solutions for a future GIL. In feasibility studies, the design of a future GIL was explained. The key technical data fixed with EDF (Electricité de France) in 1994 for the second-generation GIL is shown in Table 2.8.

The key technical data required a high-power, underground transmission system to solve network links of up to 100 km. The requirement of 3000–4000 MW power transmission is the limit for network stability reasons. Three GIL design teams (ABB, Alstom and Siemens) came up with two basic solutions: one three-phase insulated design and two single-phase insulated designs [38–41].

The three-phase insulated design had three aluminium conductors in one enclosure of steel with an aluminium inlay. The steel enclosure had the function of the pressure vessel to manage the requirements of the 15–20 bar abs. insulating gas pressure. The aluminium inlay prevented electrical shielding and allowed electrical current flow. The diameter of a three-phase GIL is about 1.5 m.

The single-phase insulated types had one aluminium conductor in one aluminium enclosure. The insulating gas pressure is about 8 bar abs. and the diameter is 0.5 m.

Both three-phase and single-phase systems have welded joints and are made to be directly buried. The gas mixture in all cases is nitrogen (N_2) and sulphur hexafluoride (SF_6). The N_2 content is between 80% and 98% and, respectively, the SF_6 content is between 2% and 20% depending on pressure and dimensions.

The three-phase and single-phase solutions have been investigated in detail to cover all aspects of manufacturing, laying process, on-site testing, operation, maintenance and repair. Three sets of prototypes have been produced by the designer to carry out all kinds of electrical, mechanical and thermal design tests.

The result was that, from a technical point of view, three-phase and single-phase GILs are both possible. Depending on the application, the three-phase GIL has an advantage in reduced space requirements but more complexity in inner conductor arrangements. The single-phase GIL needs more space for laying but is simpler in its internal conductor and insulator layout. Because of the easier laying process related to the simpler design of conductor and insulator of the single-phase GIL, this technology was chosen by Siemens and ABB for a second phase of investigation. In the second phase of investigation a prototype was made to simulate the lifetime of the equipment by laying a 100 m long GIL, directly buried. Two prototypes were built: one at the EDF test field in Les Renardières, France, close to Paris with the ABB design; the other at the IPH test field in Berlin, Germany with the Siemens design [38].

In both tests, the long-time behaviour was investigated by simulating 50 years' lifetime of the GIL in operation. The tests included the repair process and are explained in detail in Chapter 3. The prototypes had to prove on-site assembly under realistic conditions. They should contain all types of GIL elements in use. The testing requirements should cover the stresses during operation of the lifetime of the GIL.

For the tunnel-laid version a second prototype was designed and installed at the IPH test field in Berlin by Siemens. The tunnel-laid version does not need corrosion protection as with the buried version, but does need thermal expansion joints. Tunnel-laid versions of the GIL may be used in cities and metropolitan areas [42].

2.2.2.2   World-wide First GIL of Second Generation in a Tunnel

The International Telecommunication Union (ITU) chose Geneva to host the Telecom 2003 international exhibition, subject to the proviso that an additional exhibition area would be available. The City of Geneva therefore decided to construct a new hall at PALEXPO, above the A1 highway and near the Geneva–Cointrin international airport. The presence of a very high-voltage line linking Verbois to Foretaille constituted a major obstacle in the way of the future exhibition hall. This line is the property of *énergie ouest Suisse (eos)*, an important Swiss electrical utility. Different options were studied in order to find a solution for maintaining this 220 kV double-circuit line, essential to western Switzerland's very high-voltage power system and integrated in the UCTE system, linking the main substations of the Canton of Geneva [201].

Amongst the various solutions considered, it is to mention:

- It was rejected by the close airport to rise the overhead transmission line to avoid conflicts with airplanes.
- Installing aluminium pipes on the roof of the building was not acceptable because of restrictions in using the roof.
- To move the transmission line away from the PALEXPO area was not possible within the given time frame.
- The final solution chosen was an 420 m long underground tunnel. The possible transmission systems were: GIL and cables.

The GIL solution was chosen rather than another technology, because of its low magnetic interference level, its high transmission capacity and reduced heating effect, amongst other advantages. This application is the first time a GIL of the second generation has been used in connection with a high-voltage overhead line [61].

The technical data for the PALEXPO project is shown in Table 2.9.

The transmitted power of the overhead line is connected by bushings to the GIL and continued in the tunnel to underpass the newly erected exhibition hall of PALEXPO at the airport in Geneva, Switzerland.

In Figure 2.13 the principle set-up is shown.

The overhead line is connected to the GIL with the GIL disconnecting module (4) between towers 176 and 175. The GIL straight modules are laid in an S-curve with a bending radius not less than 700 m along the tunnel (2). Thermal expansion is compensated at the tunnel ends with expansion joint modules (3). The GIL is entering the tunnel at the tunnel ends by

Table 2.9 Technical data for the PALEXPO project

Type	Value
Nominal voltage	300 kV
Operation voltage	220 kV
Nominal current	2000 A
Lightning impulse voltage	1050 kV
Switching impulse voltage	850 kV
Power frequency voltage	460 kV
Rated short-time current	50 kA/1 s
Gas pressure	7 bar abs.
Insulating gas mixture	80% N_2/20% SF_6

angle modules (7). In the middle of the tunnel the GIL is fixed by a steel structure to the tunnel walls (6) and in the tunnel run held by steel structures with sliding supports (5). During erection, access was given through the installing opening (1); this was closed after erection was complete [43, 44, 61].

In Figure 2.14 a view into the tunnel is shown. The tunnel size is 2.4 m height and 2.6 m width. The pipes are held by steel structures at distances of 28 m and can slide when thermal expansion occurs. The single spiral-welded pipe has a length of 14 m.

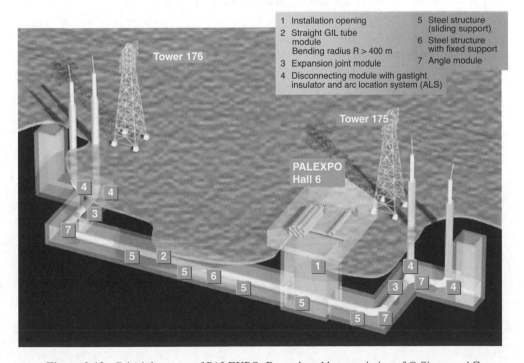

1 Installation opening
2 Straight GIL tube module
   Bending radius R > 400 m
3 Expansion joint module
4 Disconnecting module with gastight insulator and arc location system (ALS)
5 Steel structure (sliding support)
6 Steel structure with fixed support
7 Angle module

Tower 176

Tower 175

PALEXPO
Hall 6

Figure 2.13 Principle set-up of PALEXPO. Reproduced by permission of © Siemens AG

Figure 2.14   Tunnel with two GIL systems. Reproduced by permission of © Alpiq Suisse SA

The prefabricated pipes are delivered on-site into an assembly tent for preparation of the jointing process in the tunnel. The pre-assembly of the GIL sections needs to be carried out under clean surroundings to meet the requirements for high-voltage systems. Modern lightweight tents deliver such surroundings, as shown in Figure 2.15. A key element for the jointing technology is the orbital-welding machine shown in Figure 2.16.

Figure 2.15   Lightweight tent. Reproduced by permission of © Siemens AG

Figure 2.16 Orbital-welding machine. Reproduced by permission of © Siemens AG

The orbital-welding machine is highly automated and produces a constant quality of weld. The conductor and enclosure are both welded. To reach a high speed of laying it is necessary to have a highly repetitive quality level to avoid time-consuming after work and repairs of the weld. The very positive experience with this welding machine gives an optimistic view of future applications in further increasing the productivity for laying GIL over long distances. In Figure 2.14 a view into the tunnel is given on 20 January 2001, right after commissioning the system. With only 3 months of on-site work this 1 km system length of GIL was finished in record time, including a 1-week Christmas vacation for the whole crew.

The pre-assembly tent has been placed directly under the overhead line and positioned directly above the shaft connected to the tunnel right under the street. The narrow space between an airport access road on the one side and the highway to France on the other side makes it only possible to use the space directly under the overhead line for site works. The laying procedure proved to be applicable for long-distance connections. With the site experience, the productivity for assembling the GIL sections was very much increased – from two connections per shift per day to four connections per shift per day. This is a very positive experience for future projects, especially if very long distances for GIL links have to be carried out. The highly automated laying process has proven to deliver a very constant quality over the complete laying time, so that the commissioning of the system could be carried out without any failure [45, 46].

In Figure 2.17 a view into the tunnel shows how the pipes are pulled into the tunnel. The tunnel has a bending radius of about 700 m. The pipes were easily moved over the support structures with rollers and followed the bending of the tunnel [47].

In Figure 2.18 the working and laying schema of the GIL are shown for an underground tunnel. Through the tunnel opening (1) the GIL sections are brought in and then jointed by

Figure 2.17   Pulling the GIL into the tunnel. Reproduced by permission of © Siemens AG

orbital welding. The straight units (2) are brought into the tunnel, where a minimum bending radius of 700 m is allowed. The fix point (6) fixes the GIL towards the tunnel where the sliding support steel structure (5) allows thermal movement of the GIL. The expansion joint (3) takes care of the thermal movement of the enclosure. The disconnection unit (4) connects the GIL to the overhead line and separates into 500 m-long gas compartments.

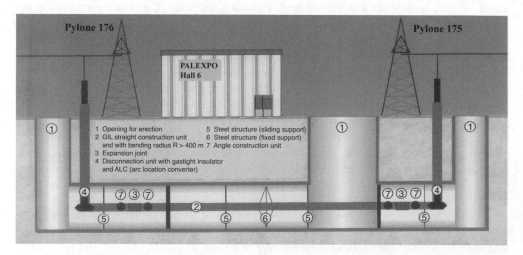

Figure 2.18   Schema of the GIL route in an underground tunnel [48]. Reproduced by permission of © Siemens AG

One of the main reasons for the customer using a GIL in this application was first, besides its competitive price, the very low magnetic field strength outside the GIL. Second the GIL is, in its electrical behaviour, similar to the overhead line and has advantages for the control and protection parameters of the complete system.

2.2.2.3 Experiences with Second-Generation GIL in a Tunnel

What was the main experience with this GIL project?

- The on-site assembly with high-voltage cleanliness (insulators, gas compartments) has shown that the conditions for high-voltage GIL can be met successfully. The automated welding process produces high-quality welds with high repetition rates. From the total of 300 orbital welds, less than 10 needed manual rework after 100% ultrasonic testing of each weld.
- The second-generation, gas-mixture GIL proved to be reliable and went into service after the high-voltage tests without any failure.
- The elastic bending down to 700 m bending radius allowed adaptation to the tunnel bending without the need for angle modules.
- The pull-in assembly of the GIL sections under bending was carried out with no problems and no time delays. The simple moving structure in the tunnel worked very well.
- The development results of the feasibility study have been proven conservative when applied in the PALEXPO project. The calculated values for thermal and mechanical stresses were below the measured values of the feasibility study which shows that the calculation parameters have been chosen conservative.

This typical GIL transmission line is part of an overhead line connection between France and Switzerland and is connected directly via bushings to the overhead line without any switching devices. The installation and commissioning of the GIL transmission line was done in a very short time without any problems. Since January 2001 the GIL has been in full operation and shown excellent power transmission behaviour and reliability. No negative impact has been detected at each power transmission level. A very long lifetime of the GIL can be expected [49].

One important reason for PALEXPO deciding on the GIL was the very low magnetic field outside the GIL, even with the rated current flowing in the conductor. The PALEXPO GIL is placed close to Geneva airport and underpasses the PALEXPO exhibition hall with its sensitive exhibitions such as the "Geneva Autosalon" or the international telecommunication fair. Therefore, the requirements of low magnetic fields were high and with a limit of 1 µT only the GIL could reach the value without expensive shielding of the tunnel. The measured value after construction of the GIL was 0.3 µT at the rated current [50].

More details on the technical design are given in Chapter 4.

2.2.2.4 World-wide First GIL of Second Generation Directly Buried

What was the need for a 400 kV gas mixture directly buried GIL?

At Frankfurt Airport an additional runway was needed to meet the increasing air traffic. North of the airport was the best location, but an existing 400 kV overhead line and substation

Table 2.10   Technical data of the Kelsterbach project, Germany

Type	Value
Nominal voltage	380 kV
Maximum voltage	420 kV
Rated current	2500 A
Lightning impulse voltage	1425 kV
Switching impulse voltage	1050 kV
Power frequency voltage	630 kV
Rated short-time current	50 kA
Rated gas pressure	8 bar abs.
Insulating gas mixture	80% N_2/20% SF_6
Power transmission	1800 MVA

were in the way. The transmission line needed to be underground with a system length of 1 km and the substation was transferred to a building using GIS [226].

The power to be transmitted was 1800 MW per system, as the line formed part of the north/south power transmission. This high power rating requires two underground cable systems per overhead line system or one GIL system. For two overhead line systems, four cable systems or two GIL systems are needed to transmit the power and a wide space would be needed. Economic reasons, but also reasons of operation, reliability and maintenance were evaluated and the result was to use transmission via a GIL.

The technical data for the directly buried Kelsterbach project is shown in Table 2.10.

The transmitted power of 1800 MVA enters the GIL at bushings, with connection points outside the airport area from the overhead line and connected to the other end with GIS at the airport. The substation connects the airport and is part of a long-distance north/south power transmission line.

In Figure 2.19 the principle set-up of the Kelsterbach project is shown.

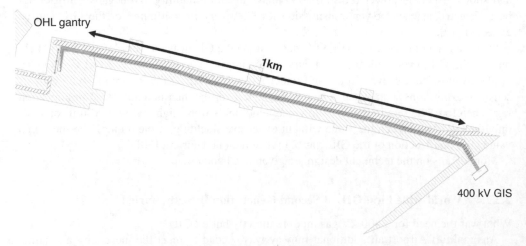

Figure 2.19   Principle set-up of the Kelsterbach project. Reproduced by permission of © Siemens AG

Figure 2.20 Bird's-eye view of the Kelsterbach project routing during construction. Reproduced by permission of © Siemens AG and © Amprion GmbH

The routing in Figure 2.19 shows that when underground solutions are chosen, nothing is straight. There are always reasons for bending and angle elements. At the overhead line gantry in the upper right corner of Figure 2.19, the two three-phase systems are collected and then enter the trench with a 90° angle element to the right. Along the 1 km route a right bend followed by a left bend form a long S shape. At the other side of the trench a 30° angle element leads the two GIL systems into the gas-insulated substation.

A good impression of the route is given in Figure 2.20.

In Figure 2.20 the trench is shown after two of the six pipes have been laid into the trench. It shows the right and left bending with a bending radius not below 400 m.

But not only right and left bendings need to be met when a GIL is laid directly buried across the country, it will also have to meet bending in vertical directions. In the end, bending of a GIL in the landscape is a spatial curve. This can be seen in Figure 2.21.

Figure 2.21 also shows the on-site construction tent, which is placed on top of the GIL trench. Inside the construction tent (in the middle of the photo) the GIL is pre-assembled in sections which are then welded to a 1 km-long transmission line of one gas compartment.

The trench has a middle depth of about 2 m where a minimum coverage of soil on the top is 1 m. Between each pipe a minimum distance of 0.5 m is needed to decouple the single phases thermally and to allow access in case of repair. The total trench width is about 6 m. The cross-section of the trench is shown in Figure 2.22.

Figure 2.21 View over the construction site at Kelsterbach. Reproduced by permission of © Siemens AG and © Amprion GmbH

The trench is backfilled with stone-free soil and directly around the GIL is covered with sand and clay for better thermal conductivity. In Figure 2.23 the six GIL pipes are shown before they are covered with backfill material.

The project in Kelsterbach shows that the on-site assembly concept, directly laid in the soil, has been successfully realized. The high-power transmission capability of 1800 MVA makes the GIL a powerful technical solution. The expected operation time is 2012. More technical design information is given in Chapter 4.

2.2.3   World-Wide Experiences

SF_6 has a three times higher electrical insulation capability than air when the same gas pressure is applied. Because of the high electronegativity of SF_6, special features of electric discharge

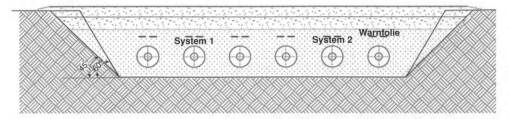

Figure 2.22 Cross-section of the trench with two GIL systems. Reproduced by permission of © Siemens AG

Figure 2.23   Trench with six pipes of GIL before backfill. Reproduced by permission of © Siemens AG and © Amprion GmbH

and the ability to attach and fix free electrons, SF_6 can be used for the design of high-voltage equipment. The compact design and small space requirements of SF_6-insulated high-voltage equipment is linked to these abilities of SF_6. SF_6 is non-flammable, inert and has high chemical and thermal stability. For humans, SF_6 is not dangerous when handled correctly, e.g. according to the IEC 62271-4 standard [51].

SF$_6$ gas-insulated, high-voltage equipment was first built in the 1960s in the USA and later in Europe. Funding equipment was installed in the 1960s and the first GIS went into operation in 1968 in Germany.

Initially, the gas-insulated equipment was made for 100 kV-rated voltage, but in short time sequences higher voltage levels were approached and SF_6-insulated equipment was installed up to 550 kV.

In 1975 the first gas-insulated transmission line using SF_6 was installed at the Cavern Hydropower Station in Wehr, at the Schluchsee, as a peak load pumping station. This GIL transmits power at 400 kV-rated voltage in the generation mode to deliver peak load electrical energy to the network and to pump water at low load times at night.

Gas-insulated transmission lines have been in world-wide use for more than 35 years in the voltage range of 145 kV up to 550 kV. Two different types of assembly are in parallel use: the flanged GIL and the welded GIL. The flanged GIL uses flanges and bolts with sealings to connect the single sections of the transmission line. The welded GIL uses orbital welding to connect these sections. Considering both systems, more than 300 km has been installed world-wide. So far, no major failure has been reported over more than 35 years of operation.

Typical applications of the GIL up to present have been links within power plants to connect the high-voltage transformer with the high-voltage switchgear, or with cavern power plants to connect high-voltage transformers in the cavern with the overhead line outside the cavern or tunnel. GIL is also used to connect a GIS with overhead lines, or as a bus duct within gas-insulated substations. These applications are carried out under different climate conditions, from the low temperatures in Canada to the high ambient temperatures in Saudi Arabia or Singapore, or the severe conditions of salty air on the coast lines of South Africa or heavy ice loads in the mountain of Europe. The GIL transmission system is independent of environmental conditions because the high-voltage system is completely sealed inside a metallic enclosure.

A second project has been constructed in Bangkok, Thailand as a connection of the 550 kV GIS with the overhead transmission line. A third project was constructed at Tehri in India inside a tunnel to connect the high-voltage power transformer at 400 kV in the cavern of a hydropower plant to the overhead line outside the mountain. A classic application. See also Chapter 7 [52].

The experiences of second-generation GIL up to now – with the PALEXPO project, Switzerland since 2001; Sai Noi, Bangkok, Thailand since 2002; and Tehri, India since 2004 – are very positive. These GIL use nitrogen and SF_6 gas mixtures as insulation medium.

In a conclusion it can be said that the experience with gas-insulated transmission lines over the last almost 40 years are very positive. The more than 300 km installed GIL over all voltage levels up to 800 kV and the current ratings up to 8000 A and 100 kA short circuit rating show that high power can be transmitted reliable with GIL. The simple basic design of aluminium pipe, epoxy resin insulators and stable, non-aging insulating gas or gas-mixtures are the fundamental reasons [52, 53, 201, 203].

It is expected that in future, with a changing transmission network towards regenerative energies, the GIL will play a major role.

3

Technology

This chapter covers various design issues: explaining the physical basics and principles of gas-insulated technology and giving reasons for each specific design criteria why GIL is best suited for high-power transmission applications even for long-distance power transmission applications.

The following questions are answered in this chapter:

What are the basic design criteria?

The basic design criteria for a GIL are explained for:

- electric field strength at rated voltages;
- maximum temperatures of the conductor and enclosure pipe;
- mechanical strength of enclosures for gas pressure and external load;
- current forces including short-circuit currents.

How to prove the functionality and quality?

To make sure that the required quality and functionality of the GIL is reached when manufactured and installed, the following tests are required and explained:

- high-voltage test
- rated current test
- short-circuit current test
- internal arc test
- mechanical tests
- vibration tests
- thermal tests and
- partial discharge test
- gas-tightness test
- resistance of conductor test.

Gas-Insulated Transmission Lines (GIL), First Edition. Hermann Koch.
© 2012 John Wiley & Sons, Ltd. Published 2012 by John Wiley & Sons, Ltd.

Table 3.1   Voltage levels

Nominal voltage levels	kV	110	220	345	380	500	735
Highest voltage of the equipment	kV	123	245	362	420	550	800
Power frequency withstand voltage	kV	230	460	520	630	710	960
Lightning impulse withstand voltage	KV	550	1050	1175	1425	1550	2100
Switching impulse withstand voltage	kV	NA	NA	950	1050	1175	1425

These tests are carried out as design or type tests to prove the correct design of the GIL. The manufacturing and installation of the GIL is proven by so-called production or routine tests. Production and routine tests are carried out in the factory or on-site and prove the correctness of the manufacturing process and the correct assembly on-site. The long-time tests are carried out to prove the quality and functionality over a simulated life. The long-time testing represents a lifetime of 50 years. The long-time test uses higher voltage and higher current levels to simulate ageing effects.

How does the GIL fit into the network?

These planning issues are investigated and studied to check the interferences of the GIL with other network equipment like cables and overhead lines when operated in parallel or in series.

The reliability of the GIL itself, and in conjunction with the network, grounding, touch voltages, safety, environmental impact and limitations, electric phase angle compensation, loadability including overloading restrictions and insulation coordination are the main issues to be covered when the functionality of the GIL at specific network conditions has to be checked.

The design of the GIL in the network depends mainly on the requirements of the power transmission rating. The power transmission rating leads to the voltage and current ratings and must fit into the power transmission network. In this chapter the focus is on the criteria and issues related to GIL to allow an evaluation of the strengths and weaknesses under certain requirements.

A comparison with other transmission technologies will give some guidelines for choosing the right technology.

The GIL is applied at all voltage levels of high voltage from 110 kV up to 800 kV. Because of the ability for high power supply, most GILs are applied at higher voltage levels of 420 kV and 550 kV. The test voltages are related to the insulation coordination for power frequency, lightning and switching impulse voltage as stated in Table 3.1.

The maximum magnetic field strength depends on the rated current and the distance from the measuring point to the GIL. The reverse current of the enclosure reduces the value below 1 μT over a few metres distance.

The GIL design follows the IEC 62271-204 "Rigid high-voltage, gas-insulated transmission lines for rated voltages of 72.5 kV and above" standard [54]. In this standard the technical requirements and test procedures are defined [55].

3.1

Gas Insulation

In Chapter 2 we showed how the gas-insulated systems for switching (GIS) and for transmission (GIL) developed over the years as part of network solutions. GIS because of its high switching capability and GIL because of its high transmission capability. Both technologies have proven very reliable in operation [56].

The short-circuit switching capability of today's equipment reaches 80 kA of short-circuit currents, and in single applications up to 100 kA. The voltage levels are up to 800 kV, with most applications at 420 kV or 550 kV. The current ratings are typically at 3000 A, and can reach 8000 A in some cases.

The gas-insulated system of the GIL is made of support insulators, gas-tight insulators, conductor and enclosure aluminium pipes and the insulating gas. The principle design criterion is the free gas space. In a GIL with a conductor pipe inside an enclosure pipe, this free space is a cylindrical sphere.

The SF_6 insulation of GIS and GIL is very well known, and many publications are available. This knowledge basis of SF_6 insulating gas is used for gas mixture insulation. When looking for alternatives to SF_6, many gases have been investigated and proven for use under high voltage conditions. But until today no other gas has found, considering insulating capability, non-toxicity and acceptable price, that could fully replace SF_6 in high-voltage switchgear and switchgear assemblies [57].

The ecological view on SF_6, with its high global warming potential, requires a closed-loop use of SF_6 from production, over-use and finally ignition into non-critical gases after use. The gas handling is defined in international standards [51, 58].

The cost aspect gives another view on the insulating gas. SF_6 is much more expensive than N_2. It offers two advantages over N_2: high arc-extinction capability and higher insulation capability.

The high arc-extinction capability of SF_6 is not needed in GIL because there is no switching function. The higher insulating capability of SF_6 can be partly compensated for N_2/SF_6 gas mixtures by increasing the pressure. This would mean that a N_2/SF_6 gas mixture GIL could be more cost effective.

The reduced arc-extinction capability of N_2/SF_6 gas-mixtures has been investigated in several studies and is also published by different institutes [59, 60]. In conclusion, the available

Gas-Insulated Transmission Lines (GIL), First Edition. Hermann Koch.
© 2012 John Wiley & Sons, Ltd. Published 2012 by John Wiley & Sons, Ltd.

knowledge about the design of N_2/SF_6 GIL was not sufficient at that time to design a reliable gas-mixture GIL. It was necessary to verify design calculations and laboratory tests with real-sized, representative test set-ups as a prototype of GIL.

The results of the investigation on the prototype test set-up are explained in Section 4.8.

3.1.1 Free Gas Space

Why is gas-insulated high-voltage equipment so successful?

The answer is down to the physical nature of gas insulation. Gases are equally distributed in a given room and gas molecules are continuously moving by gas convection. The equal distribution of gas molecules is stated by the first gas law for ideal gases: "In a given space the gas molecules of any gas or gas mixture distributes equally in the given space." For real gases the distribution will follow the Maxwell distribution in the given space which will allow difference to the ideal distribution. The impact of this differences are small and are covered by the design tolerances.

The "motor" of this movement is the partial gas pressure. This is the pressure difference between areas in the gas space. This pressure difference is driven by temperature differences in the gas. With this "motor", the gas molecules are always moving and with this movement the electrical charges are moved and mixed.

The insulating mechanism in a GIL is, therefore, similar to an overhead line where the air around the line acts as gas insulation. A GIL is like an overhead line inside an enclosure.

The advantage of free-moving gas molecules is that no electrical charges can be built up at one local spot. SF_6 as an insulating gas can capture free-moving electrons because of its electronegativity. The large design size of the GIL, with diameter about 500 mm at 550 kV voltage rating, leads to relatively low electric field strengths of typically 4–6 kV/mm as a large area mean value in the insulating gas.

In liquid insulations like oil with higher dielectric field strengths than gases, this free movement of molecules – such as gas molecules – is much slower and care must be taken to avoid particles in the oil. Because of the high relative permeability constants of oil (typically $\varepsilon_r = 5$–6), the insulation capability is higher and with this the electric field strength increases to 25–30 kV/mm. The diameter of a 420 kV oil cable is in the range of 100 mm.

In solid insulations no free movement of molecules is possible. Therefore, particles and voids of a few micrometres in size cannot be accepted in the solid insulation of the cable. The field strength in the solid insulation is down to $\varepsilon_r = 4$–5 in the range 20–30 kV/mm, and the diameter of the solid insulated cable is similar to that of oil cables.

The free gas space in the GIL between the conductor and the enclosure is cylindrical and dominant in size compared with solid insulators. More than 99.9% of the GIL is a cylindrical gas space, which offers a stable insulation and high reliability with no ageing effect. Gases do not age electrically and thermally in the range of application of GIL. The experiences world-wide with GIL show that once the GIL has successfully passed the on-site tests and is in operation at its rated voltage, the insulation will not fail.

For GIL, which is a gas-insulated system without any switching or breaking function, no major failure has been reported since its beginnings in 1974. This is good news and shows the high reliability of gas-insulated systems like GIL.

Figure 3.1.1 shows a view into a GIL.

Figure 3.1.1   View into a GIL. Reproduced by permission of © Siemens AG

3.1.2   Insulators

The conductor needs to be held in the centre of the enclosure so that the electric field distributes equally in the concentric pipe system. Insulators are typically made of epoxy resins with filler material. There are two types of filler used today: silicium (fine sand) and aluminium oxide (AlO_3). Both have the task of giving the insulator the necessary mechanical strength when moulded with epoxy resin. Several formulas for epoxy resin are used today in high-voltage gas-insulated equipment. The requirements for the various formulas come from features such as mechanical strength, maximum allowed temperature, electric insulation behaviour, surface discharge sensitivity and surface tracking withstandability.

For applications in GIL, the maximal temperature and the discharge tracking withstand-ability are the prime features, where mechanical strength and electric insulation are of minor importance.

The mechanical strength for insulators in gas-insulated switchgear is higher, where besides the gas pressure, strong shock forces from the operation of circuit breakers and switches also need to be covered.

The electrical insulation capability of the insulating material is not fully used because the dimensioning (diameter of the enclosure) is given by free gas space design of the insulating gas.

The insulating capability of epoxy resin used in GIL is in the range of $\varepsilon_r = 3\text{--}5$ compared to gases. This means, the other way around, that the insulator is operated 3–5 times below its physical insulation capability; a high safety margin. This design feature of GIL is another reason for the high reliability in operation.

There are three types of insulator in use today:

- Post-type insulators with one, two or three legs holding the conductor in the centre of the enclosure.
- Conical insulators, which are concentric around the conductor with holes and fix the conductor towards the enclosure.
- Gas-tight conical insulators, which are concentric around the conductor and enclosure without holes to separate the gas compartments and to fix the conductor towards the enclosure. They are part of the pressurized enclosure of the GIL.

3.1.3   Gas-Tight Enclosure

Why is gas-tightness so important for GIL?

There are two reasons for the high gas-tightness of GIL. First, there should be no loss of gas in the system because it is needed to keep the insulation of the high voltage and to stay in operation. Second, if there is no gas loss, then there is also no impact from the environment to the high-voltage system inside. If nothing is coming out, nothing will go in. Besides dust and all kinds of particles, the low level moisture is important for the high-voltage insulation. Insulating gases need dew points of $-20°C$ and below at atmospheric pressure.

A dew point of $-20°C$ makes the insulating gas very dry. Moisture would manage to enter the gas compartment even if there is an overpressure in the GIL. Moisture in the gas would decrease the insulation capability and an internal discharge would be the consequence. Therefore, high gas-tightness is required over the 50-year expected lifetime. This gas-tightness is reached with O-ring sealings and welded joints. On-site welding is tested 100% by ultrasonic weld control, which proves the gas-tightness.

Automated orbital-welding machines produce multilevel welds to joint the enclosure pipes. Depending on the wall thickness, 5 to 10 layers of welds form the joint and avoid voids in the aluminium which could cause gas leakage. In Figure 3.1.2 a cross-section of such a multilayer weld is shown. In Figure 3.1.3 the set-up of the automatic orbital arc-welding process is shown.

The enclosure material is sheet aluminium or extruded aluminium. To increase the mechanical stability aluminium alloys are used. Both materials are gas-tight because of their molecular

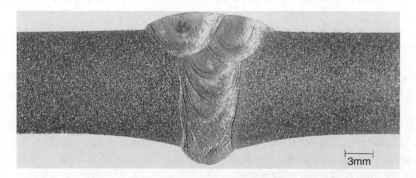

Figure 3.1.2   Cross-section of a multilayer weld. Reproduced by permission of © Siemens AG

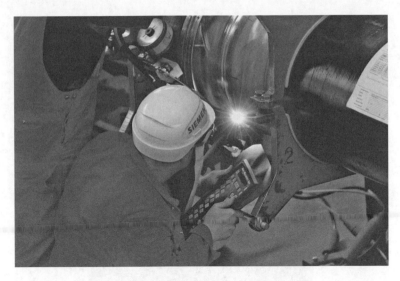

Figure 3.1.3 Orbital arc-welding process. Reproduced by permission of © Siemens AG

structure. The sheet aluminium is produced by milling under high pressure from high material thickness to sheet strengths of typically 5–10 mm. Under the high pressure of milling the aluminium forms a homogeneous structure and is gas-tight. Sheet aluminium is formed into pipes by milling, and the plates or coils are connected by longitudinal welding (Figure 3.1.4) or by spiral welding (Figure 3.1.5).

Figure 3.1.4 Longitudinal welded pipe. Reproduced by permission of © Siemens AG

Figure 3.1.5   Spiral welded pipe. Reproduced by permission of © Alpiq Suisse SA

The extruded aluminium pipes are pressured from a block of aluminium through a pressure tool into a seamless pipe. The structure of the material is longitudinal and homogeneous and delivers gas-tight pipes.

3.1.4   Insulating Gases

What is the right insulating gas?

The first gas used for high-voltage insulation was ambient air with any air insulated system. When metallic enclosed high-voltage systems were developed, technical air (N_2 with O_2) came to be used to make the gas more unique. Inert gases such as helium (He) or argon (Ag) followed as additives to improve the insulation capability, but they are expensive and difficult to handle. In 1920 the artificial gas sulphur hexafluoride (SF_6) was designed for chemical use, mainly as a chemical reaction stopper by excluding oxygen from the mould. In the 1960s the electrical properties of insulation and electrical arc distinguishing were used to design the first electrical systems. In GIL, no switching operation is necessary because the purpose of the GIL is electric power transmission only. Therefore, only the insulation capability of SF_6 is used in GIL [60].

SF_6 is an expensive gas compared with nitrogen (N_2) or dry air, and in GIL large volumes of insulating gas are needed. For a length of 1 km of GIL under 8 bar pressure, 500 m^3 of gas is needed. The cost, and the need for insulation alone and not switching, led to the use of an SF_6/N_2 gas mixture with a majority of 80% N_2 and a minority of 20% SF_6.

The use of N_2 in combination with SF_6 already has a long history, back to the 1970s. In switchgear for very low-temperature applications, SF_6/N_2 gas mixtures were used to reduce the liquefaction point to temperature values below $-30°C$. N_2 is, like SF_6, an inert gas with low chemical reaction potential and, therefore, good to use in high-voltage equipment. The gas mixture of SF_6 and N_2 shows, even at small percentages of SF_6 (e.g., 10–20%), good high-voltage insulation properties. One reason is the character of pure SF_6 for the corona-stabilizing effect. This corona-stabilizing effect explains the ability of SF_6 to capture free-moving electrons and to form a corona cloud around a strong inhomogeneous field, e.g. a pike or scratch on the conductor surface. With this corona cloud the SF_6 heals the failure. This means that electrons which are emitted into the gas at inhomogeneous field areas of high electric field strength are captured by the SF_6 molecule, which send out a photon. The number of photon are seen as the a corona cloud. Inhomogeneous field areas are usually small failures e.g. a scratch on the surface of the conductor which creates a high electric field strength to send out these electrons. Of course there are limits to healing failure; surface quality in high-voltage systems is high.

The complexity of the SF_6 molecule with a number of energy levels and its electronegativity, which lead to the corona-stabilizing feature are the main reasons why SF_6 is used in GIL and GIS and offers high reliability for lifetime operation. Because, in the real world of manufacturing and installation, ideal surfaces and shapes can only be reached in given tolerances, SF_6 as an insulating gas helps to handle quality tolerances in a high-voltage system.

Dry air or technical air (N_2, O_2) would be another practical option for insulating gas mixtures. Compared with SF_6 a loss of insulation capability needs to be accepted, which finally leads to larger sizes or lower voltage ratings by about 50%. Many other gases could be seen as gas mixtures, with or without SF_6. Inert gases like helium or argon are one family of gases, but they are even more expensive than SF_6 and have less insulation and additionally no arc-extinction capability. Chlorinated hydrocarbon (CHC) is also a possibility, but in practical use under high electric fields strength it may produce carbon strings in the gas (carbon fallout), which is not acceptable in high-voltage equipment. These carbon strings can cause a flashover in the GIL.

This leaves SF_6, N_2 and dry or technical air as practical possibilities. When SF_6 is mixed with N_2 the relationship between insulating capability and percentage of SF_6 and N_2 is shown later in Figure 3.1.6. A mixture of dry air (75% N_2, 23% O_2, 2% other gases) or technical air (80% N_2, 20% O_2) would show similar results to the N_2/SF_6 gas mixture, but with a lower level of absolute insulation capability and in consequence a larger dimensioning. That is why these gas mixtures are not used widely today [61].

3.1.4.1   Sulphur hexafluoride SF_6

The insulating gas SF_6 offers two basic advantages for use at high voltage:

- high insulating capability because of strong electronegativity;
- high arc-extinction capability because of recombination and arc cooling.

For GIL, the arc-extinction property is not needed, therefore, a mixture with nitrogen (N_2) is possible without losing too much of the insulation capability. A 20% SF_6 content in the 80% N_2 majority is chosen, as explained in Section 3.1.4.3.

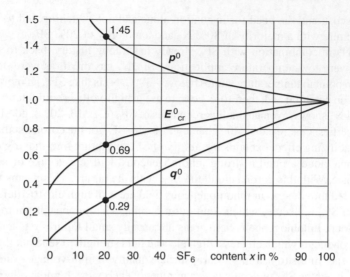

Figure 3.1.6 Normalized properties of N_2/SF_6 gas mixtures. Reproduced by permission of © Technische Universität München

The chemical/physical properties of SF_6 are as follows:

- non-flammable
- colourless and odourless
- non-toxic
- not soluble in water
- five times heavier than air
- high chemical and thermal stability
- high heat storage capability allows use only in closed pressure systems.

Electrical properties of SF_6:

- high insulation capability
- high arc-extinction capability.

Sulphur hexafluoride is a technical gas which has a high global warming potential (GWP) and can only be used under specific gas-handling procedures, as explained in international standards and in the European Union regulations under the F-gas regulation [51, 58].

3.1.4.2 N_2

Nitrogen is chemically stable as the N_2 molecule. N_2 is an inert gas, which means it is not a very active molecule to react chemically with other molecules. N_2 has a no electronegativity, which means that free electrons cannot be bound with the molecule, easy only at very low energy levels. N_2 is a widely available gas and can easily be produced out of air. Therefore, the price is very low. Compared to SF_6 the molecule size of N_2 is three times smaller and

N_2 has no electronegativity, which means it can not capture electrons like SF_6. In case of a inhomogeneous field strength electrons are getting accelerated through the gas and can hardly be stopped by N_2. The small size leaves much free space in between so that electrons can find the way through the gas and the missing electronegativity can not catch electrons in the gas and send out an photon. This task, of stopping and binding free electrons, can be done much better with SF_6. The large molecule of one sulphur and six fluorine atoms blocks the free-moving part of the electron, taking out the kinetic energy, binding the electron and sending out a photon. The photons we can see form the so-called corona, a light shining effect around the high-voltage conductors in high electric field areas.

N_2 has a good partner in SF_6 to design the economical high-voltage power transmission system GIL. The lower electrical insulation capability of N_2 leads, in practical use at high voltage, to very high gas pressures. For this reason, the N_2 insulated systems are limited to about 100 kV in practical use. Higher voltage levels of 420 kV or 550 kV would require large diameters of 1.0 m and a high gas pressure of 15 bar to 20 bar compared with a diameter of 0.5 m when a 20% SF_6/80% N_2 gas mixture is used at 0.8 MPa. These large diameters and high pressures would make the GIL design costly, and for AC applications pure N_2 gas insulation has not been used at these higher voltage levels.

3.1.4.3 N_2/SF_6 Gas Mixture

3.1.4.3.1 Principle of Gas Mixture

Much research has been done since the early 1970s on replacing SF_6. Some of the tested gas mixtures showed higher insulation capabilities than SF_6, but all of them are more critical in terms of toxicity or degenerating under partial discharge initiated by high electrical field strength or switching arcs, so that the wide use of such gas mixtures in high-voltage equipment is not recommended. A summary of the research work has been presented in several international publications [62–69].

It is shown that there is today no alternative to SF_6 for switching purposes [59, 73], but that there is an alternative to pure SF_6 for insulation purposes: a mixture of nitrogen (N_2) with only a small amount of SF_6 added. GIL need a large amount of gas only for insulation purposes, so this gas mixture of mainly nitrogen with as low as possible a content of SF_6 is recommended. To fix the percentage of SF_6, an optimization process is needed to find the best ratio between SF_6 content, gas pressure and enclosure diameter. The characteristics of N_2/SF_6 gas mixtures show that with an SF_6 content of less than 20%, an insulating capability of 70–80% of pure SF_6 can be reached at the same gas pressure [70, 218].

Insulation Capabilities
The investigations made since the 1970s indicate that nitrogen and sulphur hexafluoride can be used for high-voltage applications. The strong increase in voltage withstandability with only a small percentage of SF_6 supports the goal of reducing the total amount of SF_6 used in a long-distance GIL [70–73].

The dimensioning of the GIL with N_2/SF_6 gas mixture is based on realistic test set-ups. The limiting high-voltage impulse withstand voltages and the related withstand electric field strengths under the limiting conditions of N_2/SF_6 gas mixture and gas pressure are defined.

For the measurements, a coaxial cylindrical field of a 500 mm enclosure pipe and a 120/190 mm conductor pipe was used. The pressure range covers values of 0.1 MPa to 0.7 MPa. Gas mixtures with 0% to 20% SF_6 have been investigated.

The lightning impulse voltage level depends on the gas pressure and for different gas mixture relations of SF_6 and N_2 shows a non-linear behaviour. When the gas pressure is increased, the non-linear behaviour of N_2/SF_6 gas mixtures is the same as for pure SF_6. For switching impulse voltage the same tendencies are seen. Based on this curve, the gas pressures of different gas mixtures of N_2 and SF_6 can be chosen to reach the same level of withstand voltage. The withstand voltage for pure SF_6 at 0.3 MPa is reached at 0.5 MPa for gas mixtures of 20% SF_6 and at 0.7 MPa for gas mixtures of 10% SF_6. These pressure values are the minimum operating gas pressure. The maximum gas pressure according to pressure vessel standards is 0.7–0.8 MPa [70].

The pure N_2 insulation would require a gas pressure of 2.0 MPa, which would need a very costly pressure vessel design. The pure N_2 solution is not seen as a practical solution, but will be investigated in the gas mixture studies as an orientation value.

The measurements of pure N_2 show large variations of power frequency, switching and impulse voltages. Only 5% SF_6 content of the N_2/SF_6 gas mixture delivers much more stable test results, which are similar to pure SF_6 in behaviour [70].

Best Point Design
In SF_6/N_2 gas mixtures with 20% SF_6 and 80% N_2, the absolute insulation capability of 80% from pure SF_6 is reached. This means that with only 20% of the volume of SF_6, a GIL can be designed at moderate gas pressure of 0.8 MPa and acceptable diameter of 500 mm for 420 kV.

When the SF_6 content of the gas mixture is reduced, the design point for the GIL will move into the step area of the voltage withstandability curve. This means a small change of SF_6 content (e.g., from 5% to 6%) would cause a large change in absolute high-voltage insulation capability. Not a good point for the basic design.

When the SF_6 content reaches the 20% mark, the curve is flat and a small change (e.g., from 20% to 21%) does not cause a large change in the absolute insulation capability of the gas mixture. This is the reason why in practice the SF_6 content is kept around 20%.

N_2/SF_6 gas mixtures have been investigated in long-term measurements using real-sized test set-ups in parallel with the test laboratory of the High Voltage Institute of the Technical University of Munich and at the High Voltage Test Laboratory of Siemens in Berlin. The measurements show that the principle behaviour of the N_2/SF_6 gas mixture follows the good electrical characteristics of pure SF_6, including the corona-stabilizing effect. The measurements also show that even a small percentage of SF_6 will increase the absolute insulation capability strongly. With an SF_6 content of 20% in the N_2/SF_6 gas mixture, more than 70% of the insulation capability of pure SF_6 is reached.

Normalized Properties
The properties and advantages of N_2/SF_6 gas insulation have been under investigation for more than 40 years. The use of N_2/SF_6 gas mixtures has been restricted to low-temperature applications to avoid liquefaction of SF_6 in operation at temperatures down to $-50°C$. The first real application came with the second-generation GIL. The main driver was the cost reduction for long-distance applications.

The handling of the N_2/SF_6 gas mixture needs a normalized evaluation tool which allows us to find the right pressure, mixture ratio and volume [61].

Already at low SF_6 concentrations, the N_2/SF_6 mixture acquires high dielectric strength due to the unique synergetic physical properties of these two components. The normalized properties of N_2 and SF_6 gas mixtures are shown in Figure 3.1.6.

Figure 3.1.6 gives three curves for the dielectric strength of N_2/SF_6 gas mixture depending on the 100% SF_6 value. The diagram shows the following.

- Normalized intrinsic dielectric strength E_{cr}°. The E_{cr}° value for pure N_2 is below 0.4, while a 20% SF_6 content improves the value to 0.69. The pure SF_6 value is 1.
- Normalized pressure p^0 required for equal dielectric strength. The p^0 value for the 20% SF_6 content of the gas mixture is 1.45, which means the pressure is 45% higher compared with pure SF_6. The p^0 curve increases strongly with lower SF_6 content and will reach a factor of about 3 for pure N_2.
- Normalized quantity q^0 of SF_6. The curve for q^0 shows an almost linear increase of the quantity of SF_6 in the gas mixture. At 20% SF_6 a value of 0.29 is shown, which means only 29% of the SF_6 quantity is needed at this gas mixture compared with pure SF_6. Only a moderate pressure increase of 40% is necessary in order to arrive at the same critical field strength as pure SF_6, but the amount of SF_6 in such a mixture is 70% less.

N_2/SF_6 mixtures have been regarded for some time as a potential alternative to pure SF_6 and much relevant research work has been performed. But it is mainly fundamental physical data and properties ascertained in small test set-ups under ideal conditions that have been published. Many further investigations were therefore necessary to achieve the specified assured insulation capability according to the established requirements for reliable operational service of a real GIL.

Internal Arc Pressure Increase
The pressure increase due to an internal arc is, in N_2/SF_6 mixtures, higher than in pure SF_6 of the same dielectric strength. But the gas volume of the GIL design is extremely large. The final overpressure is therefore restricted to values far below the critical value for fragmentation. The arc cross-section is, in N_2/SF_6 mixtures, much larger than in pure SF_6 and the thermal power flow density into the enclosure at the arc root is therefore much less. There is therefore very small erosion at the enclosure in the event of an internal arc in such a gas mixture. Even during autoreclosure functions there will be no puncture of the enclosure or any external effect in the event of an internal failure in the GIL [61].

The N_2/SF_6 gas mixtures have been used in high-voltage switchgear for many years in extreme low-temperature regions around the world with very positive experiences [74].

Moisture
In gas-insulated systems, moisture plays a major role in reliable insulating capabilities. Dry gases, with a dew point below $-20°C$, are really dry when used in high-voltage systems. In the presence of moisture, the insulation capability of the gas will be reduced and the GIL will not reach the full voltage level.

To avoid moisture in gas compartments, the gases used are industrially dried and kept this way in bottles. The insulating materials (like cast resin insulators) are also kept in a dry condition after production, with short times between manufacture and use in the gas-insulated system. This reduces the moisture penetration into the material. The gas compartment is dried before filling with insulating gas by using N_2 for pressure tests.

The sealing system of gas-insulated systems needs to be gas-tight, which is clear, but it also needs to be moisture-tight. Because of the high difference between ambient moisture in air and the dry insulating gas, a moisture "pressure" is always present around the seal. Today, high-quality O-ring sealing takes care of the need for dry insulating gases.

When joints are welded, the weld needs to be moisture-proof, which is given when the weld is gas-tight. Gas-tightness of welded joints is measured by a 100% ultrasonic test and in case of a detected void by additional X-ray weld inspection. To be sure that the moisture in the gas-insulated system stays low for the total lifetime, drying materials are added to the gas compartments.

3.1.4.3.2 Verification Tests

To verify the dielectric properties of the gas mixture, several prototype test set-ups have been realized at real size and conditions.

The prototype test set-up is a real-sized test set-up which includes all the required GIL elements to fulfil the function of electric power transmission under high voltage. The goals of the investigation with the prototype test set-up are:

- N_2/SF_6 gas mixture relationship
- gas pressure
- surfaces of insulators, conductor and enclosure
- test voltages
- potential low-grade inhomogeneity.

Typical dimensions of the coaxial cylindrical electric field are 400 and 500 mm enclosure pipes and 120 mm, respectively 190 mm, conductor pipes to form a free gas space to avoid the impact of the non-coaxial cylindrical electric field conditions at the ends and at insulator locations.

The test set-up elements have standard manufacturing quality and realistic surface structures. The general requirements of the power frequency, switch and lightning impulse test voltages are shown in Table 3.1.1.

Not only ideal measurements have been carried out on the test prototypes, but also the impact of failures has been investigated using the partial discharge measuring technique.

Table 3.1.1 Rated withstand voltages

Lightning impulse withstand voltage	1425 kV at 18% SF_6
	1300 kV at 13% SF_6
Switching impulse withstand voltage	1050 kV at 18% SF_6
	950 kV at 13% SF_6
Power frequency withstand voltage (r.m.s. value)	630 kV at 18% SF_6
	520 kV at 13% SF_6

The partial discharge measurements with specific built-in known defects as peaks on the conductor and free-moving particles show for the N_2/SF_6 gas mixtures with low SF_6 content only small differences from the partial discharge measurements in pure SF_6. In the amplitude/phase angle/intensity diagram, practically no differences have been detected. This means that the UHF partial discharge measuring method of pure SF_6 can also be used for N_2/SF_6 gas mixtures. The high-voltage testing showed that no basic differences can be seen with phase measurements, comparing N_2/SF_6 gas mixtures with pure SF_6 [75, 76].

Measuring Equipment
Different prototypes of new developments of GIL using N_2/SF_6 gas mixtures successfully underwent high-voltage testing. Monitoring light and sound emission during testing with voltages higher than the rated withstand voltages helped significantly to determine the location of disruptive discharges. High voltage testing on site can include fast rise impulse voltages. For typical test sections with a length of about 1 km, the minimum useful front duration has been calculated to be about 30 µs.

Prototype 1
Initiated by a research and development programme of Electricité de France (EDF) in 1994, the first phase towards the GIL of the second generation was concluded with so-called lab prototypes. Basic considerations and high-voltage tests led to the choice of N_2/SF_6 mixtures with a high content of nitrogen as insulating gas, and an optimal enclosure diameter of 650 mm for a power transmission capability of 3000 MVA for the directly buried GIL. See Figure 3.1.7.

Figure 3.1.7 Prototype test set-up at EDF Laboratory in France. Reproduced by permission of © Alain Sabot

Prototype 2

The second prototype test set-up for gas mixture investigation has been installed twice. Once at the high-voltage laboratory at Siemens Switchgear Factory in Berlin, Germany and then again at the IPH test laboratory, also in Berlin, Germany.

The gas mixture chosen was an N_2/SF_6 mixture with a high content of nitrogen and as low as possible a content of SF_6. As there are a lot of interdependencies between all design parameters, an optimization process was established to find the optimum between SF_6 content/gas pressure/enclosure diameter and design temperature. The basic behaviour of N_2/SF_6 gas mixtures shows that with an SF_6 content of less than 20%, an insulating capability of 70–80% of pure SF_6 can be reached with a slightly higher gas pressure. In Figure 3.1.8 the test set-up is shown to verify the design of the GIL under high-voltage test conditions. The set-up includes all basic elements and is approximately 30 m long. The IEC ratings for the test voltages have been applied for power frequency, switching and lightning impulse, and the results have proven the reliability of the system.

The test arrangement in Figure 3.1.8 consists of a disconnecting unit in the front, left side of the photo, connecting to two 12 m-long straight units, and at the end a high-voltage test bushing connected to the straight units by an angle unit. This real-sized test set-up of 30 m length is needed to verify the applied test voltages under real service conditions.

The impact of the different ratios of gas mixtures of SF_6 and N_2 to the insulators, conductors, contacts and sliding contacts was investigated to prove the gas mixture reliability and to find the right pressure and gas mixture ratio.

Figure 3.1.8 High-voltage test of straight units. Reproduced by permission of © Siemens AG

Considering the present-day criteria of ecology and economy, questions are again being raised for alternatives. N_2 can be looked upon as being absolutely uncritical in terms of environmental compatibility because, as a naturally occurring gas, it is encountered in large proportions in the atmosphere. N_2 is therefore suitable as a basis for insulation. Pure N_2 insulation, though, would result in unrealistic designs for the required insulation levels. But even by adding some SF_6 an insulating capability is achieved which, from the present point of view, is realistic in relation to practical use.

In terms of arc-quenching capabilities, a substantial reduction must be expected with an N_2/SF_6 gas mixture, with the result that, considering requirements arising from constantly increasing short-circuit capacities, such a mixture is not feasible for use in circuit breakers. That is why the following ignores aspects of arc-quenching properties in circuit breakers [61].

The analyses of the insulating capabilities of N_2 and N_2/SF_6 mixtures undertaken by many institutes in the 1970s and 1980s provided a large number of clues for evaluation. The 50% breakdown voltages for various voltage applications were generally determined on small test set-ups. The results indicate trends and peculiarities which, however, are not always relevant or applicable to large-scale installations [59].

A variation of the test set-up is shown in Figure 3.1.9. Here, in addition to the straight units, at the end an angle element with a 90° directional change to the left is added before the connection to the bushings.

The alternative test set-up including an angle element is representative of a real directional change in GIL projects. The angle unit with its 90° directional change has an impact on high-frequency transients reflected at the angle unit and superposed on the test voltage during lightning impulse voltage testing and with lower frequency for switching impulse voltage testing.

Prototype 3

A third prototype was installed at the University of Munich and at Siemens Switchgear Factory in Berlin, both in the high-voltage test laboratories.

Different from prototypes 1 and 2, these test set-ups had a much smaller gas compartment. Large gas compartments need long handling times to evacuate the volume down to some millibars and to refill the volume. Prototype 3 was made to study the gas mixtures with real-sized isolators in large measuring series to get statistical security.

To determine design principles, the breakdown and withstand voltages were determined on basic test set-ups. The first test arrangements essentially consisted of coaxial cylindrical fields (diameters 120 mm/400 mm and 190 mm/400 mm) with a pure gas section (Figure 3.1.10) to first rule out possible influences of solid insulators on the results.

The analyses were undertaken in the pressure range from 0.1 to 0.7 MPa. The effects of the SF_6 amount in N_2 were analysed up to 20% volume. The two ranges were chosen in advance because they are realistic in relation to a gas mixture GIS or GIL design without entering into unrealistically large dimensions. The pressure range embraces the pressures that are usual for current GIS installations, with related designs for the pressure resistance of the housings. The strong dependence of the insulation strength from pressure and SF_6 amount in the analysed gas mixture range offers a wide leeway for the geometric design of GIS or GIL assemblies.

Therefore, tests on real-sized arrangements are necessary for the design of GIL with a "new" insulating medium, with the aim of achieving an assured insulating capability for the corresponding conditions under real live operation [34].

Figure 3.1.9   High-voltage test of straight units including angle unit. Reproduced by permission of ©
Siemens AG

Test Procedures
The gas withstand capabilities were investigated with 50 Hz AC voltage and standard switching
and lightning impulse voltage in both polarities. An assured withstand voltage was determined
in each case, i.e. a reproducible maximum possible withstand voltage without breakdown.

During the course of the AC voltage tests the voltage was increased several times up to the
breakdown voltage, and the withstand voltage that can be achieved just below the breakdown
voltage was verified. In principle, the procedure for the determination of the maximum impulse
withstand voltage was similar (see Figure 3.1.11).

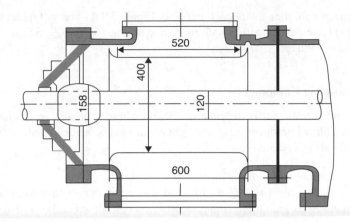

Figure 3.1.10 Test set-up for investigation of gas mixtures. Reproduced by permission of © Siemens AG

The voltage was increased in steps up to a breakdown. After the first breakdown, it was attempted to achieve a withstand voltage with 15 impulses without breakdown that was 3% less than the breakdown value. If further breakdowns occurred, the voltage was reduced by a further 3% step. The ultimately achieved withstand value (15 impulses without breakdown) was referred to as the assured maximum withstand value. The waiting time between the impulse voltages applied was taken as some minutes to let the gas stabilize again and to have

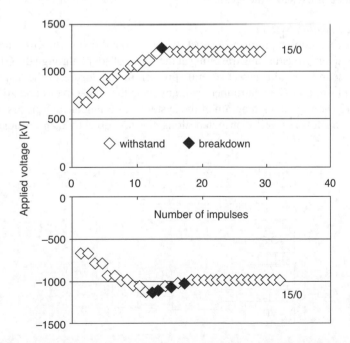

Figure 3.1.11 Example of determination of a positive and negative maximum withstand switching impulse voltage (15 impulses without flashover). Reproduced by permission of © Siemens AG

equal measurement conditions with each impulse. Figure 3.1.11 shows typical sequences for this withstand voltage determination with reference to a test with 10% SF$_6$ at 0.6 MPa with switching impulse voltage.

3.1.4.3.3   Measuring Equipment

During high-voltage testing it is important to know where a disruptive discharge is taking place. A gas breakdown produces intensive light and a loud acoustic signal. Both phenomena can be used in locating systems.

Light Detection through Inspection Window
The following system – consisting of readily available equipment – enables the pictures from four video cameras to be monitored simultaneously on one screen and recorded on one video recorder (Figure 3.1.12).

Each camera feeds into an electrical/optical converter. Camera, converter and battery are situated in a metallic housing (C1 to C4). Optical cables transmit the video signals to optical/electrical converters (O/E), which are connected to a quad compressor (Q). A video recorder (VR, with reverse single-frame capability), a monitor (M) and a video printer (P) complete the instrumentation. The control room should be shielded so that standard (comparatively inexpensive) equipment can be used.

Prerequisite for this system is that the video cameras and quad compressor are able to detect all sparks. This can be checked by directing all cameras at one spark gap and producing one spark; all cameras must show the spark.

Acoustic Bang Locating System
Each insulation breakdown produces a shock wave of sound inside the GIL enclosure which can be detected as an acoustic bang travelling at the speed of sound through the GIL. Therefore, measuring the sound arrival times at several spots on the enclosure of the test object allows the breakdown location to be determined by comparing the respective sound arrival times.

Figure 3.1.13 shows a block diagram of the system: several acoustic sensors are connected to the metallic enclosure, well distributed along the test object. Each encapsulated sensor

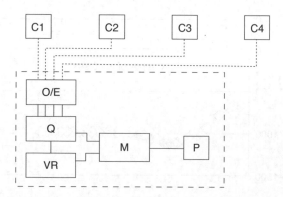

Figure 3.1.12   Block diagram of video monitoring system. Reproduced by permission of © Siemens AG

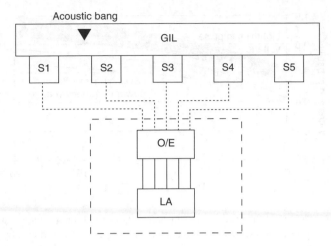

Figure 3.1.13 Block diagram of acoustic bang location system. Reproduced by permission of © Siemens AG

(S1 to S5) is battery powered, and the only output is a binary light signal if a certain sound level is exceeded. Signal transmission via optical fibres through an optical/electrical converter (O/E) to a logic analyser (LA) provides electromagnetic compatibility for the system. The logic analyser and any equipment which may be connected for storage and documentation should be situated in an electromagnetically shielded room.

The first signal triggers all binary channels. The sound velocity in the enclosure material is a few thousand metres per second, many orders of magnitude slower than electromagnetic (including optical) wave propagation; therefore, the lengths of the optical cables do not influence the result.

A typical logic analyser record (Figure 3.1.14) provides some details:

- The acoustic bang occurred close to sensor 2 (compare the mark ▼ in Figure 3.1.13); the step triggered all traces of the logic analyser at the instant T_T.
- The later the steps of the other sensors occur, the longer the distance between the discharge and the corresponding sensor.

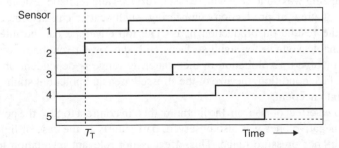

Figure 3.1.14 Pattern for a acoustic bang as indicated by the mark ▼ in Figure 3.1.13. Reproduced by permission of © Siemens AG

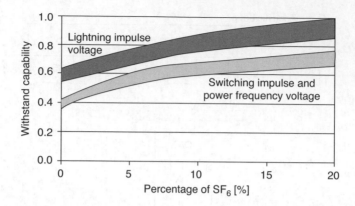

Figure 3.1.15 Comparison of the maximum withstand values of lightning impulse, switching impulse and AC voltage for N_2 insulation with small amounts of SF_6 admixed. Reproduced by permission of © Siemens AG

This system can easily be checked at the beginning of the tests by hitting the enclosure with a hammer. It can be calibrated after a disruptive discharge, because hitting the enclosure close to the discharge location produces a similar logic analyser pattern as the original acoustic bang.

3.1.4.3.4   Test Results

The corresponding withstand field strengths were calculated from the determined assured maximum withstand voltages as a basis for new designs. Figure 3.1.15 shows the comparison between the different voltages (lightning impulse, switching impulse and AC voltage) of the withstand field strengths determined as a function of the amount of admixed SF_6 at a total pressure of 0.6 MPa.

Conspicuous about the results is that no significant difference was recognizable between the switching impulse values and the peak values of the AC voltage, with the result that both voltage types could be combined in one scattering range.

With the large number of single values over the parameters of pressure and SF_6 amount involved, the deviations underlying the individual measured values result in a corresponding range of scatter. The condition of the test set-ups makes an essential contribution to the cause of the scatter. As testing was done here with realistic surfaces and realistic residual contamination from the point of view of production engineering, such scatter can also be expected when implementing the results on production parts. The top envelope can be interpreted as the maximum attainable voltage for small surfaces with the lowest roughness and without any contamination involved. Exploitation of the scattering range is dependent on the respective application. Surface condition, the size of the stressed area and potential slight contamination must be taken into account.

The well-known effect of N_2 from the literature that, in contrast to SF_6, the positive values lie lower than the negative ones was also observed, particularly in the case of lightning impulse voltage in the higher pressure range. This effect is not relevant in relation to fundamental design tasks, though, because evidence has to be provided for both polarities. This is why there is no need for a differentiated consideration [59].

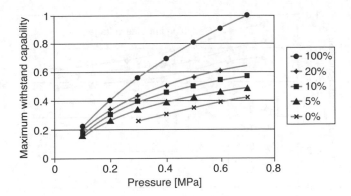

Figure 3.1.16 **Figure 3.1.16** Maximum withstand capability of various N_2/SF_6 mixtures in comparison with pure SF_6 for lightning impulse voltage of pure SF_6. Reproduced by permission of © Siemens AG

Comparison between N_2/SF_6 and SF_6

The pressure dependence of the withstand capability of the gas mixtures is not so distinct in the case of N_2/SF_6 as for pure SF_6. Figure 3.1.16 shows the maximum withstand capabilities for lightning impulse voltage in the analysed pressure range for various mixing ratios in comparison with SF_6. The SF_6 values were determined in the same way as in the case of the gas mixture analyses described above. Their applicability to GIS constructions has proven itself in many years of practice. The non-linear response to pressure increase, as is known with SF_6, is evident in the same way in the case of the various gas mixtures. The same trends can be observed in the case of switching impulse voltage.

Figures 3.1.17 amd 3.1.18 show the reduction in maximum withstand capability of the mixtures to pure SF_6 for lightning and switching impulses. As the pressure increases, the relative withstand capabilities, which is the ratio of the withstand capability of N_2/SF_6 gas

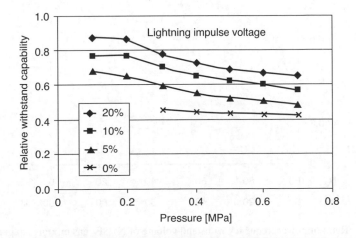

Figure 3.1.17 Ratio of the withstand capability of N_2/SF_6 gas mixture to pure SF_6 for lightning impulse voltage. Reproduced by permission of © Siemens AG

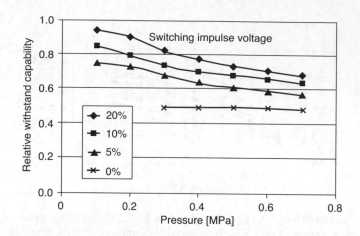

Figure 3.1.18 Ratio of the withstand capability of N_2/SF_6 mixtures to pure SF_6 for switching impulse voltage. Reproduced by permission of © Siemens AG

mixtures to pure SF_6 for lightning impulse voltage, decrease continuously in the case of the gas mixtures in comparison with SF_6. In the case of lightning impulse voltage, the reduction is slightly more distinct than in the case of the switching impulse voltage. Accordingly, a pressure increase leads to a disproportionately larger system design. When pure N_2 is used, this trend is not recognizable, although a certain amount of inaccuracy has to be considered here due to a too small number of measured values.

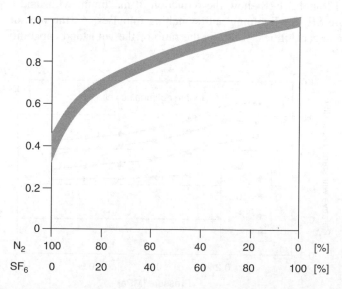

Figure 3.1.19 Ratio of power frequency withstand voltage of N_2/SF_6 gas mixtures to pure SF_6. Reproduced by permission of © Siemens AG

Practical Application of Gas Mixture

New electrode geometries can be designed and optimized with a knowledge of the maximum attainable withstand field strength and with powerful field calculation programs. For GIS and GIL SF_6 installations, this fundamental approach has been practiced for years and has also proven itself in the design of 420 kV GIL with N_2/SF_6 gas mixture insulation. This has been verified in tests of individual elements, in the type tests and in the qualification test for GIL in conformity with IEC 62271-204 [54].

The high-voltage type tests were performed on approximately 30 m GIL segments containing all possible elements of the respective GIL system (Figures 3.1.7–3.1.9). The long-duration qualification test embraced a large number of tests on 70 and 100 m-long GIL segments (see Section 3.8) that had been fitted under on-site conditions and were tested successfully. As proof of system safety, the high-voltage testing accompanied all works and tests done on the qualification set-up, and confirmed the design. With gas mixture insulation, the technology of gas-insulated lines, which with SF_6 installation has been known for many years, has evolved positively [76].

For the use of a gas mixture insulation consisting of N_2 with small amounts of SF_6, a large number of measurements were undertaken on real technical set-ups. A wide spectrum of mixing ratios and pressures was covered by the tests. The maximum dielectric capability was derived from the tests carried out, and the basis for the design of GIS and GIL components with N_2/SF_6 gas mixture insulation was elaborated. No fundamentally deviating phenomena were discovered in comparison with the approach towards a design using pure SF_6. The correctness of the determined design variables was confirmed in a large number of tests on 420 kV GIL components and segments and also on GIS assemblies.

A simplified curve is given in Figure 3.1.19. The width of the curve covers the static bandwidth and shows the ratio of power frequency withstand voltage of N_2/SF_6 gas mixture to pure SF_6. It gives simple information about the impact of SF_6/N_2 gas mixtures.

3.2

Basic Design

What is basic design? Basic design of a GIL gives the basic answers for the layout: maximum electric field strength, maximum gas pressure, type of insulating gas and maximum current rating. The answer to these basic design criteria is the rated voltage and current, rated pressure and pipe diameters and wall thickness. With basic design also the length of each GIL section, the size of the gas compartment, the type of joints and the type of laying is explained.

3.2.1 Overview

Comparing GIL Technology with GIS

Most of the gas-insulated technology of the GIL is directly taken from the GIS. In comparison with GIL the GIS is much more complex with the functionality of circuit breaker, disconnector, ground switch, ct, vt and bus bars. The GIL is defined only for power transmission without any switching element or instrument transformer. The closest component to a GIL is the bus bar of the GIS. Even here sometimes ground switches are integrated in the bus bar.

From a dielectric position the GIL can be explained as a coaxial configuration with only a few inhomogeneous locations (the insulators). Higher field strengths are linked to locations of insulators. The big majority of GIL are quasi homogeneous, coaxial, cylindrical electrical fields. The dimensioning of the GIL is designed for insulator locations. Compared with homogeneous field strength, the field strength value is only 60% of the required value in the homogeneous part of the GIL.

Two main differences between GIL and GIS have to be noted: the use of N_2/SF_6 gas mixtures and the long length of GIL. The use of N_2/SF_6 gas mixtures reduces the absolute insulation capability by about 50%, which is compensated by the size and increased gas pressure. The long length of GIL has an impact on surge impedance transients in the system, with reflections at the open or closed ends. This needs to be covered by surge arresters [77].

Influencing Factors

There are many factors to recognize in the development to cover all the different aspects of the GIL in use. The main factors which influence, directly or indirectly, the dimensioning of this high-voltage technical solution are explained in the following: insulation coordination is linked to the rated voltage of the equipment. When the rated voltage is fixed, the test voltages

Gas-Insulated Transmission Lines (GIL), First Edition. Hermann Koch.
© 2012 John Wiley & Sons, Ltd. Published 2012 by John Wiley & Sons, Ltd.

can be found depending on the network conditions. Network conditions depend on the type and size of power generation, the transmission network technology (overhead lines, cables, GIL) and the environmental requirements (lightning, humidity, salty air).

Overvoltages

IEC and IEEE give orientation in choosing the right test voltage levels for power frequency, lightning and switching impulse test voltages. Mostly, overhead lines are used in the transmission network of today. Overhead lines are subject to lightning strikes and related transient overvoltages. These overvoltages depend on the type of tower structure and the geographical situation in the area where the towers are standing. This has an impact on the amplitude and probability of transient overvoltages and the related test voltages.

Using overvoltage surge arresters to reduce the voltage level of transient overvoltages is an opportunity which shall be investigated in any case. Using integrated protection by surge arresters, which means direct coupling of the arresters to the GIL gas compartment at each end and/or at defined positions along the route, the transient overvoltage level can be limited which leads to lower insulation levels of the GIL. The use of surge arresters reduces the diameter of the GIL and the volume of insulating gas. This has a great impact on cost reductions.

Current Rating

Besides the rated voltage, the rated current has a major impact on the design of the GIL dimensioning. The current-carrying capability depends on the cross-section of the conductor pipe, which depends on the pipe diameter and the wall thickness. The maximum current-carrying capability is defined by the maximum allowed temperature. The temperature limit is given by the maximum allowed temperature of the insulator material at the location where the insulator is contacted to the conductor pipe. The maximum temperature is defined with a safety difference and, therefore, the thermal dimensioning of the GIL leads to relatively low temperatures at the enclosure pipe accessible from outside. The large size of the enclosure pipe (e.g., 500 mm for a 550 kV GIL) offers a large surface between the encloser pipe and the soil to transport the heat coming from the conductor as a source to the ambient air in a tunnel or to the surrounding soil when directly buried. In a tunnel, the heat will be transported by convection of the tunnel air and in the soil, conductivity transports the heat.

The dimensioning of the enclosure and conductor pipe of the GIL depends on heat dissipation of the GIL, depending on the condition in the tunnel or in soil when directly buried. The final design diameter of the pipes may be larger than the requirement coming from the high-voltage insulation, when heat dissipation has higher requirements.

Another design parameter is the short-circuit current requirement. The higher the short-circuit rating, the higher the mechanical and thermal requirements of the GIL. The short-circuit requirements come from the network and the power available at the point where the GIL is connected. In principle, the GIL offers very good short-circuit behaviour when it is single-phase insulated. In this case the electrodynamic forces are concentric to the conductor and do not give additional mechanical stress to the system. The material and design of the insulators have to meet these requirements. These mechanical requirements are contradictory to the dielectric requirement, where the optimum is to use as little as possible insulating material, which makes the insulator mechanically weak. A typical two-parameter optimization process.

Additionally to this two-parameter optimization, the impact of gas pressure, soil load, thermal expansion, bending or mechanical forces at sloops turns the two-parameter optimization into a multi-parameter optimization.

The high expectation in lifetime, of 50 years or more, comes from the high cost of such a power transmission investment. No ageing effects for the GIL under full operation and high-voltage application is known for the insulators, the enclosure or conductor pipe or the insulating gas when operated in the defined application range. From this point of view the GIL does not really have a limitation in lifetime. The period of 50 years is chosen based on the 40 years of experience with GIL.

N_2/SF_6 Gas Mixture
One main aspect for the development of the second-generation GIL was the reduction of SF_6 used in the GIL by introducing the N_2/SF_6 gas mixture with 80% N_2 and 20% SF_6. Two goals are reached with this: reduction of the cost of insulating gas and reduction of the ecological impact. N_2 is available for much lower prices than SF_6, and with the application over long distances the volume gets larger and therefore the cost reduction effect is larger, too. The SF_6 is always kept inside the gas compartment and not released to the atmosphere at any time, by proper handling processes as explained in international standards [51, 58, 204, 205].

Two Concentric Aluminium Pipes
The enclosure pipe is a pressurized compartment using aluminium alloys. In addition, the mechanical load of bending – and in case of directly buried pipe, the soil load – has to be added. The enclosure pipe also carries the induced reverse current as a consequence of the induction low absolute current level of the current in the conductor. The conductor pipe is manufactured of pure electrical aluminium to keep the conductivity high and the transmission loss low. The conductor pipe is centred in the middle of the enclosure pipe and is held by insulators. Three types of epoxy resin insulators are used: post-type insulators for conductor support, conical insulators for conductor fixing towards the enclosure and gas-tight conical insulators to form independent gas compartments [78].

Sliding Contact System
A sliding contact system consisting of a plug and socket compensates the thermal expansion of the conductor pipe between two conductor fixing points at about 100 m distance. In between the fixing points of the conductor pipe, each 10–15 m in length, a post-type insulator supports the conductor.

Orbital Welding Joints
The GIL segments of 10–15 m length are orbital welded to joint together with an automated welding process. The welding seams are 100% quality checked using automated, orbital ultrasonic test equipment, a technology developed primarily to prove the pipe joints in the radioactive section of a nuclear power plant. The total jointing process offers very high quality of each joint and is reproducible in large numbers because of its automated design.

Operational Parameters
The operational parameters of the GIL explain the electrical and mechanical operation conditions of the GIL. These parameters are the basis of calculating the transient and continuous transmission, the inductive and capacitive impact to transmission in a network, the switching on and off impact, the transient overvoltage and other conditions of net stability. In Table 3.2.1, typical values of technical parameters for two GIL types are presented for example. For each

Table 3.2.1 Technical parameters

Value	Unit	GIL 1 Ø 500 mm	GIL 2 Ø 600 mm
Resistance R'	mΩ/km	9.42	6.94
Reactance X'	mΩ/km	6.75	6.48
Inductance L'	mH/km	0.215	0.206
Capacitance C'	nF/km	54.45	57.08
Reactive power $U_n = 400$ kV Q'_C	MVar/km	2.74	2.87
Surge impedance Z_W	Ω	63.1	60.3
Natural power $U_n = 400$ kV S_{nat}	MVA	2536	2869

project the impact of the GIL under the given network conditions needs to be calculated to prove the network stability for the different network conditions [79].

3.2.2 Dielectric Dimensioning

The dielectric dimensioning of gas-insulated systems follows maximum allowed electric field strength and is proven by test voltages. The typical field strength to be allowed is about 20 kV/mm, and is the design criterion for internal minimum radii of bench and edges. The value of 20 kV/mm can be higher if the affected area is not too large; this is very much related to manufacturer experiences and the type of material chosen [80].

The test voltages to prove the dielectric dimensioning are high-frequency lightning impulse voltages with a rise time of 1.2 μs and an impulse backflange of 50 μs. These high-frequency test voltages are more sensitive for peak-like inhomogeneous areas. Switching impulse test voltages with a rise time of 250 μs are used for article detection. The 50 Hz or 60 Hz power frequency test voltage is more sensitive for large-area electric fields.

The dielectric dimensioning is basically optimized by the best relationship in a cylindrical system of the inner conductor and outer enclosure pipe when the logarithm of the diameter relationship is one. This means that the outer diameter is 2.73 times larger than the inner diameter 2.73 relates to the number 'e' [81].

3.2.3 Thermal Dimensioning

The thermal requirements coming from the current ratings and the inner and outer temperatures are defined by the wall thickness of the enclosure and conductor pipes. Typical limiting temperatures of the insulators are 105–120°C, with maximum allowed enclosure pipe temperatures of 40 or 50°C for directly buried GIL and 60 or 70°C for tunnel-laid GIL. Each project has its own specific parameters for the thermal laying and, therefore, a specific layout of the wall thickness. This also has a great cost impact, because each millimetre of wall thickness adds cost [54, 82].

With such large test set-ups the correct dielectric design is verified to guarantee a reliable operation in service.

3.2.4 Insulation Coordination

Besides the dielectric dimensioning of the GIL itself, an additional possibility of adapting the equipment to network requirements is provided by using overvoltage surge arresters.

The integrated surge arrester reduces the maximum values of overvoltages in the GIL and, therefore, allows a lower insulation level. The metal oxide (MOX) arrester used needs to be coupled directly to gas compartments of the GIL in order to also protect against high-frequency transient overvoltages.

Because of this integrated surge arrester the transient overvoltage stress of the GIL can be reduced compared with an overhead line. The definition of the basic insulation level (BIL) depends on many influencing factors, such as the MOX arrester, characteristic, soil resistances and ground wire on the overhead lines.

The GIL can be seen as a closed electrical system with integrated transient overvoltage surge arrester at each end or in several sections of the transmission line. No external electrical impact is given to the GIL because of its metallic and solidly grounded enclosure. Compared with an overhead line system with a BIL of 1425 kV, the BIL of a GIL can be reduced to 1300 kV. This means that the dielectric dimensioning of the GIL can be reduced. Further reductions of BIL for long-distance transmission are possible when integrated surge arresters are placed along the transmission line [91].

3.2.5 Electrical Optimization

Electrical optimization is a process whereby the dielectric design and insulation coordination are fixed after the mechanical parameters have been chosen to optimize the overall electrical behaviour. Therefore, detailed electric field calculations have been carried out to design the insulators and to reach a design of homogeneous field distribution in the GIL as far as possible.

To prove the calculation using real-sized test set ups it was decided to have the measurements verified at two independent test set-ups. Original parts were used, to be as realistic as possible. A full test programme covering power frequency, lightning and switching impulse test voltage at gas pressures up to 2.0 MPa and with various SF_6/N_2 gas mixture ratios was carried out. The parallel measurements at two locations in two test set-ups excluded systematic failures in the test programme. To avoid systematic failures in the measurements two equivalent test set-ups have been installed in Berlin and Munich.

In conclusion, it can be said that the dielectric strength of the gas mixture GIL – including effects of particles and surface effects of the insulators – is 5–10%, for a minimum of SF_6 in the gas mixture. Starting at this percentage level, the SF_6 shows its corona-stabilizing feature which guarantees reliable function and long lifetime. The gas pressure of 1.0 MPa should not be exceeded to limit the impact of inhomogeneous field areas on the conductor. In the following, the details are explained [81].

3.2.6 Transmission Network Studies

To coordinate the dielectric design with the requirement of the network it is necessary to calculate the system behaviour with network calculations. In these network calculations, the different network conditions are simulated. Typically, the following situations are simulated using the dielectric parameters of the GIL: short circuit between phases and towards ground, switching the line with open end and lightning strikes in the connected overhead line near to or far from the GIL connection point. This requirement is a system and network requirement and is explained in detail in Chapter 4.

3.2.7 Gas Pressure Dimensions

The GIL is filled with an N_2/SF_6 low-pressure gas mixture. The typical filling pressure is about 0.8 MPa and is seen as a low-pressure vessel. In addition to the low gas pressure, the electrical insulating gases used are non-corrosive and extremely dry. Inner corrosion can therefore be excluded. This is the basis for special and adapted pressure vessel standards for use in high-voltage equipment only. These standards, with a dew point of $-20°C$, have been published by CENELEC as European standards and are generally accepted world-wide.

Depending on the type of material and the manufacturing process, the following standards are used:

- EN 50052 Cast aluminium alloy enclosures for gas-filled high-voltage switchgear and controlgear [83].
- EN 50064 Wrought aluminium and aluminium alloy enclosures for gas-filled high-voltage switchgear and controlgear [84].
- EN 50068 Wrought steel enclosures for gas-filled high-voltage switchgear and controlgear [85].
- EN 50069 Welded composite enclosures of cast and wrought aluminium alloys for gas-filled high-voltage switchgear and controlgear [86].
- EN 50089 Cast resin partitions for metal enclosed gas-filled high-voltage switchgear and controlgear [87].
- EN 50087 Gas-filled compartments for a.c. switchgear and controlgear for rated voltages above 1 kV and up to and including 52 kV [88].

The operational experiences of gas-insulated systems have shown that insulating gas pressures in the range of 0.4 to 0.8 MPa provide high reliability. When the gas pressure is increased to 1.5 to 2.0 MPa, the dielectric sensitivity towards internal small failures and areas of inhomogeneous electric fields gets higher. Any gas at high pressure, in principle, is more sensible when fast transient voltages are applied.

From a mechanical point of view the pressure vessel cannot be seen as low pressure any more, and more effort is needed to design such enclosures. For these reasons today most gas-insulated systems are operated at a gas pressure of about 0.8 MPa or below.

3.2.8 High-Voltage Design Tests

High-voltage design tests give the requirements for the equipment, coming from the high-voltage network as defined in the subsection "Insulation Coordination". The design test values are given by the related standards IEC 62271-204 [54], IEC 62271-1 [82] for each voltage level of the network. In Table 3.2.2 the test voltages are shown.

For the 400 kV and 500 kV network, the maximum voltage U_m is 420 kV and 550 kV for the equipment during operation. Based on this value, with a factor of 1.5 for $U_m = 420$ kV and 1.36 for $U_m = 550$ kV, the power frequency withstand voltage is given. Behind these factors a statistical evaluation of the possibility of an internal flashover is given. Because the background is static for impulse voltages, statistical flashovers during the design test are allowed. Most standards require 15 impulse tests and allow 3 flashovers.

Table 3.2.2 Dielectric-type test values

Test parameters GIL		
Nominal voltage	400 kV	500 kV
Maximum voltage (U_m)	420 kV	550 kV
Power frequency withstand voltage (U_{PF}) 1 min	630 kV	750 kV
Lightning impulse voltage (U_{LI})	1425 kV	1600 kV
Switching impulse test (U_{SI})	1050 kV	1200 kV

The ratio of lightning impulse voltage to the rated voltage is 3.39 for $U_m = 420$ kV and 2.9 for $U_m = 550$ kV. For switching impulse voltages the ratio is 2.5 for $U_m = 420$ kV and 2.18 for $U_m - 550$ kV. These test voltage ratios are standard values and reflect the experiences made over the last 40 years. It also shows that the safety margin at higher voltage levels are smaller which has economical background in reducing the size of the equipment. Lightning and switching impulses are required to prove the design for transient voltages, which may come from lightning strikes in connected overhead lines, and for switching operations in the network.

In Figure 3.2.1 the test set-up at the IPH high-voltage test facility from 1998 is shown. At the front of Figure 3.2.1, a straight GIL unit is shown. On the right, the test bushing is connected by two disconnector units. In the middle of this straight section a compensator unit

Figure 3.2.1 High-voltage test set-up at IPH, Berlin, Germany. Reproduced by permission of © Siemens AG

Figure 3.2.2   High-voltage test set-up at EDF, Les Renardières, France. Reproduced by permission of © Alain Sabot

is integrated. At the left end of the test hall an angle unit makes a 30° turn to the left. The air bushing for the high-voltage test is connected to the power frequency test voltage generator in the middle of the photo or by the impulse test voltage generator behind. The total length of the set-up is about 25 m.

In Figure 3.2.2 the test set-up for the high-voltage tests at EDF, Les Renardières, France in 1995 is shown. The total length of the test set-up is about 40 m. At the left end of the GIL a bushing connects the disconnecting unit with the high-voltage test equipment. In the middle of the picture two 12 m-long straight units of GIL are seen, which are separated by a compensation unit to compensate for thermal expansion. At the right end of the test set-up another disconnecting unit separates the gas compartments from the left two-thirds length to the right one-third length. Finally, at the very right-hand end a disconnector unit concludes the high-voltage test set-up.

3.2.9   Current Rating Design

The current-carrying capability of the GIL is mainly dependent on the conductor cross-section and the ambient thermal conditions. The high current rating follows strongly the requirements coming from the specific project. Each project will need a thermal layout to choose the right diameter and wall thickness of the GIL.

In principle, the GIL offers a high current rating capability because of its relatively large diameter (e.g., 500 mm). This large diameter requires a minimum pipe wall thickness starting

Figure 3.2.3   High-current test set-up. Reproduced by permission of © Siemens AG

at about 6 mm. For the 500 mm enclosure pipe the active cross-section is about 10 000 mm^2 and for a 180 mm diameter conductor pipe it is about 3300 mm^2. The maximum wall thickness of enclosure pipes is 10 mm and for conductor pipes 16 mm, and the cross-sections are increased accordingly.

The second important aspect for the current rating is related to the ambient conditions in the tunnel or when directly buried in the soil. These parameters depend on the project conditions and need to be calculated in advance to find the right design of the GIL in terms of diameter and wall thickness. For large scale projects of e.g. 100 km system length this means that the GIL will be designed exactly to the project requirements. Not so for smaller project e.g. 100 m, here standard design values for the diameter and wall thickness will be used.

In Figure 3.2.3 the design test set-up for the current ratings is shown. Each of the GIL elements is included in the test set-up: straight unit, angle unit, compensator unit and disconnecting unit.

3.2.10   Short-Circuit Rating Design

The short-circuit rating for the GIL is given by the requirements of the network. Depending on the network location, different levels of short-circuit ratings are given when close or far away from the power plant. The closer the power plant, the higher the short-circuit rating. In some cases in the network several power plants are connected together in one transmission network knot. Typical values of short-circuit ratings are found in the transmission network: 50 kA, 63 kA and 80 kA. In some rare cases 100 kA can be found.

The short-circuit rating has two requirements on GIL:

• To withstand the electromagnetic forces between conductors and ground.
• To limit the maximum conductor temperature because of the extreme high current.

In a single-phase concentric system as GIL, with a solidly grounded enclosure, the electromagnetic forces are also concentric around the conductor and centralize the conductor in the enclosure. The insulators, therefore, do not need to withstand high mechanical forces. The solidly grounded system causes an induced current in the enclosure in the opposite direction. Both forces are substrated and the resulting forces are small. The thermal heat-up of the system because of the high current rating is negligibly low because the large cross-sections of the enclosure and conductor pipe can compensate the heat and only a very low temperature rise can be measured. This is related to the short time of the short-circuit current, of 0.5 s or less. The aluminium conductor and enclosure pipes of the GIL act as heat storage for this short time. The required tests are carried out with the same test set-up as the current rating test in Figure 3.2.3.

In case of a short circuit the currents can be very high. In a 400 kV transmission network the typical ratings are 50 kA or 63 kA or 80 kA. Such high currents produce heat and strong electromagnetic forces. In the situation of a short circuit the dynamic current at a 63 kA location in the network can reach 170 kA for the first cycle of the net frequency. This high current causes high magnetic forces between the conductor and ground and towards the other phases. The insulators and conductor pipes need to withstand these forces.

In a single-phase GIL system which is solidly grounded, the magnetic forces are concentric around the conductor and focus the conductor in the centre. This takes forces away from the insulators. The single-phase concentric GIL has very good short-circuit force withstandability.

Different from a three-phase system – where the full forces have to be taken care of between the conductors or in case of cross-bonding – only a small portion of the forces are superposed and the reduction of forces is low.

3.2.11 Internal Arc Design

The internal arc design is a safety value for withstandability in case of failure when an internal arc may occur. The reason for an internal arc could a failure of the insulator which can cause an arc at the surface of or inside the insulator. An other reason might come from particles inside the gas compartment or from failure at the conductor or enclosure pipes which can cause high electric field strength and an flash over in through the insulating gas. The internal arc is the worst case of failure that can happen to a GIL. The very high power rating and operation in the transmission network where high power is available at any point of the network makes it necessary to design a safe technology which does not give high risk to operators or public.

The public is usually excluded, as the GIL is installed in a tunnel or directly buried in the ground. The operators need to have access to the system for operation and maintenance checks [89].

When an electrical arc is initiated inside a gas compartment, the hot temperature will heat up the gas and the pressure in the gas will increase. This pressure increase depends on the height of the current and the volume of gas. The smaller the gas compartment, the higher the gas pressure. That is why small gas compartments use pressure eruption discs which break open when a defined pressure is reached and release the gas pressure.

In GIL, the gas compartments are very large; up to 1 km-long gas compartments are possible and the pressure increase of a short-circuit-rated arc current is low and does not exceed allowed pressure values. Therefore, no gas eruption disc is needed in a GIL.

The second criterion for safe GIL is that the internal arc does not burn a hole into the enclosure wall, which is under pressure of about 0.8 MPa. The physical behaviour of the arc helps here. In 80% N_2/20% SF_6 gas mixtures the arc footprint is, like in air, relatively large. This large area results in low footprint temperatures and a low level of melting of the enclosure aluminium wall. No burn-through of the wall results.

The internal arc test is really a hard test for the equipment, which can be better understood when the following comparison is made. The short-circuit arc current of, for example, 63 kA is driven by an arc voltage of about 8000 V. This results in a power of about 500 MVA, which is the power of a mid-sized power plant. The energy in 0.5 s is then 250 MW, which is 69 kWh. This is about 248 MJ, which can heat 1000 l of water from room temperature (20°C) to boiling point (100°C).

For the development and design test several test set-ups including all major components of the final GIL design has been chosen to prove the arc withstand capabilities of the enclosure in case of an internal arc. According to IEC requirements, a typical arrangement has been chosen. The length of the arrangement was about 25 m, so that the gas pressure rise, because of the heat of the arc, stayed within acceptable limits. In future installations of the GIL, gas compartments will be large, so the gas pressure rise during an internal arc fault will not be significant. The set-up has been assembled in the lab, and for safety purposes rupture discs have been foreseen but should not open during the test. Figure 3.2.3 shows the arrangement for an internal arc test.

Summary of the test results:

- No external influence during and after the internal arc test has been noticed.
- No burn-through of the enclosure has occurred/very low material consumption has been noticed on the enclosure and conductor.
- The pressure rise within the enclosures during arcing was significantly low, so that even the rupture discs did not open.
- The arc characteristic is much smoother in moving speed and wider arc discharge canal compared with the characteristics in pure SF_6 (e.g., wider arc diameter/lower arc travelling speed).
- Cast resin insulators are not seriously affected.

All these results speak in favour of the safe operation of the GIL – even in the very unlikely case of an internal arc, the environment is not affected. The results of the arc fault test also showed that in case of a tunnel-laid GIL, no personal danger to people in the tunnel could be expected.

The operational safety of the metallic encapsulated GIL is tested by the extreme situation of an internal arc to protect the operation personnel which can come close to the GIL in operation.

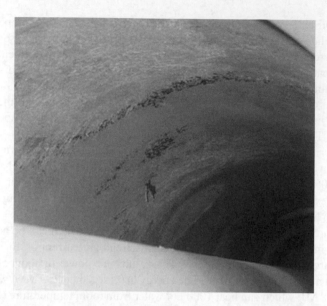

Figure 3.2.4   Result of an internal arc. Reproduced by permission of © Siemens AG

In Figure 3.2.4 the result of the arc test is shown, with very little impact to the enclosure wall even at an arc current of 63 kA for 0.3 s duration.

The photo shows that only a few micrometres of aluminium material have been melted by the internal arc. No burn-through or rupturing of the enclosure can be seen.

3.2.12   Electromagnetic Current Forces Design

A set-up containing the most sensitive parts against electrodynamic forces has been tested in the IPH test labs, Berlin. The chosen Z-arrangement containing two 90° angle units and an additional expansion below has been tested under short-circuit stresses. With this Z-arrangement, maximum mechanical stress has been applied to the GIL, especially to the insulators. The test set-up for these tests is the same as for the current rating tests and internal arc test (see Figure 3.2.5).

Summary of the test results:

- All the components have withstood the electrodynamic forces initiated by a simulated short-circuit current.
- The resistance of the main circuit (conductor and enclosure) has not been increased after the test (in fact, it has been slightly decreased).
- All components were in functional condition after the test (including the sliding contacts).

3.2.13   Mechanical Design

The mechanical design of the GIL depends on the type of laying: tunnel-laid or directly buried. The elastic bending of the GIL also gives stress to the enclosure and conductor pipes, but bending is needed to follow the landscape contours without getting directional change elements like elbows.

Figure 3.2.5 Z-arrangement for testing the electromechanical forces. Reproduced by permission of ©
Siemens AG

Tunnel-Laid GIL
For tunnel-laid GIL the relevant pressure vessel codes are applicable and give design rules for
the wall thickness of the enclosure pipes. In addition, the mechanical forces in the enclosure
material coming from elastic bending need to be covered [38].

Directly Buried GIL
When the GIL is buried directly in the soil, additional to the mechanical forces in a tunnel the
mechanical forces of thermal expansion and soil load have to be covered. These additional
mechanical forces are covered by using stronger aluminium alloy. With finite element method
(FEM) calculation, the GIL has been simulated for all operational situations to optimize the
mechanical design. In Figure 3.2.6 the FEM calculation of a directly buried angle element is
shown [90].

3.2.14 Integrated Overvoltage Protection

The GIL allows a direct connection of surge arresters inside the GIL gas compartment. Surge
arresters are of the same type as already used for GIS. Surge arresters may be connected to
the disconnecting housing and can be positioned at each point along the transmission line.

This integrated surge arrester protects the GIL from overvoltages which can occur by
lightning strikes to the overhead line. The integrated surge arrester is part of the GIL gas
compartment and directly coupled to the conductor. This means a low surge impedance for
the coupling of the surge arrester and therefore an effective protection also for high-frequency
overvoltages. The surge arresters can also be connected via air bushing to the overhead line
directly at the GIL bushing.

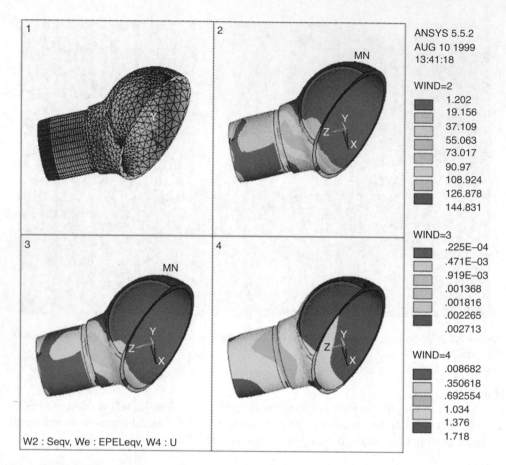

Figure 3.2.6 Finite element calculation of a GIL angle element or directly buried laying. Reproduced by permission of © Siemens AG

The surge arresters are positioned at both ends of the GIL and reduce overvoltage stresses. In some cases additional surge arresters may be positioned along the transmission line to reduce the overvoltages further. The calculations in Völcker and Koch [91] show possible reductions of overvoltage stress values that can be reached in a GIL by external and integrated surge arresters.

3.2.15 Particles

The impact of particles on the dielectric dimensioning is large. Particles in gas-insulated systems may be free moving, conducted to the insulators or to the conductors. Many investigations have been carried out over the last decades to understand the impact of and the technical solutions to particle problems [92].

The assembly of high-voltage gas-insulated systems needs to be done under clean room conditions to avoid particles entering the high-voltage gas compartment. This quality control process of the assembly is essential and can be shown in GIS factories or at GIL assembly tents on-site.

Cleanliness is one aspect of keeping particles away from high-voltage equipment. The second important aspect is to allow safe areas for particles inside the high-voltage system. Such safe areas are defined by low electric field areas, like a manhole in GIS. Free-moving particles are lifted by the electromagnetic field of the high-voltage electrostatic force. This electrostatic force of 50 Hz power frequency will force the particles to move at 100 Hz mechanical forces. The particles will dance.

When lifted, the force will lead the particles to the highest electrical field – which is the conductor. With a voltage of zero after 10 ms, the particles will fall down towards the lowest electrical field at the enclosure.

If there is a manhole, for example, or any other "electrical field shadow" in the form of a particle trap then the particles will go there. Once lying in the manhole or particle trap, the high-voltage electric field will not be strong enough to bring the particles back. The particles have gone out of the field of high voltage and will not cause any danger to the system. This process is part of the commissioning of the GIL, and is called high-voltage conditioning. The GIL is a self-cleaning system, if the process of cleaning is used correctly [93–95].

3.2.16   Thermal Design

3.2.16.1   General

The thermal design of the GIL is a basic dimensioning criterion for the GIL. The generated heat of the current in the conductor and the induced current in the enclosure needs to be transported to the environment around the GIL. This may be surrounded by air, or in case of buried GIL the soil around the GIL. In this section it can only be shown in principle how the heat flow is calculated in the case of a GIL directly buried in the soil or laid in air, as in a tunnel or above ground. In detail, it is shown how the internal heat of the conductor is transferred to the enclosure [78, 90, 96, 97].

3.2.16.2   Heat Transfer Inside the GIL

General
Gas flow caused by natural convection between two horizontal isothermic cylinders has been investigated by many authors before. Such investigations are typical for solar collectors, heat-storage installations and pressurized vessels of reactors. In this case the gas flow inside the GIL has been investigated. Many experiments, test set-ups and numerical calculations have been made to study laminar or turbulent gas flow for low Rayleigh numbers (up to 107) and for isothermic cylinders. Only a few studies are known which look into convection at non-isothermic conditions and at high Rayleigh numbers (greater than 107). In this experimental and theoretical study the differences of the gas flow formulas and their dependence on the Rayleigh number, the Prandtl number* and the relationship of radii are shown here. The results show the temperature distribution and the local heat transformer coefficients [52, 96, 98].

---

*Rayleigh and Prandtl numbers are fluid related constants for numerical calculations.

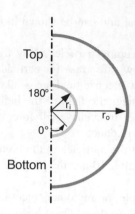

Figure 3.2.7   Scheme of the ring area. Reproduced by permission of © Siemens AG

Task of the Study
The study investigates the numeric solution of a stationary, two-dimensional natural convection
in a ring area between two horizontal concentric cylinders (see Figure 3.2.7). The inner ring r_i is
under constant heat flow caused by the electrical current and has a constant temperature T_i. The
outer ring r_o has a constant lower temperature T_o. The thermo-physical behaviour of the fluid
is taken as constant, except the fluid density in the upstream area of the moment equation. The
Rayleigh number is in the region of 10^6 to 10^{10} and the radii relation is 2.5 to 2.6. The gravity
force is vertical to the bottom. A schematic of the physical model is shown in Figure 3.2.7.

Limiting Conditions
To solve the five equations, limiting conditions for the gas flow and the walls are needed.
In different case studies the inner cylinder is heated up at constant flux and kept at a constant
temperature. The outer cylinder is at a constant temperature T_o. The geometry is a vertical
symmetry level, the solution is made for the vertical half level. The axial and vertical speed
are zero along the wall. At the symmetry line the axial speed and the axial heat transfer are
zero. Additionally, in case of constant heat transfer between the surfaces of the two cylinders,
radiation is used.

Results and Discussion
The radii relationship $R+ = 2.5$ and the Prandtl number $Pr = 0.77$ are almost constant [52].
      The heat transfer and gas flow lines are shown in Figure 3.2.8 for the gap between the inner
and outer cylinder. The gas flow lines start at the inner conductor, which heats up the gas. On
the top, the heated gas is flowing vertical to the outer enclosure and flows along the cylinder of
the outer enclosure back to the bottom of the cylindrical gap. During this down flow the heat
of the gas is transferred to the enclosure and the cooled gas sinks down to the bottom of the
cylindrical gap. From there the cycle starts again by heating up the gas flow at the conductor
surface.
      This convective gas cooling system is the physical background reason for the good heat
transfer from the conductor pipe to the enclosure pipe and from there to the surrounding
atmosphere or soil.

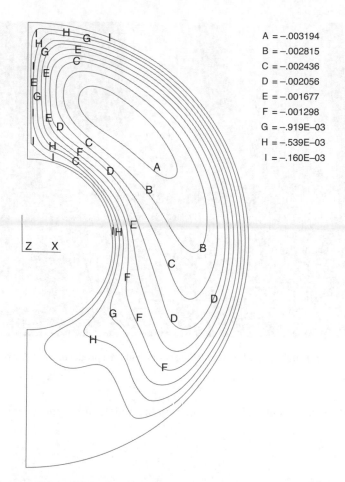

A = −.003194
B = −.002815
C = −.002436
D = −.002056
E = −.001677
F = −.001298
G = −.919E−03
H = −.539E−03
I = −.160E−03

Figure 3.2.8   Gas flow lines. Reproduced by permission of © Siemens AG

Experimental Thermal Test Set-up
The GIL in a thermal test set-up is shown in Figure 3.2.9. The inner and outer cylinders of
the GIL are made of aluminium alloys. The outer diameter of the inner cylinder is 250 mm
and the wall thickness is 16 mm. The inner diameter of the outer cylinder is 630 mm and the
wall thickness is 10 mm. Between both cylinders a gas mixture of SF_6 and N_2 is used as an
insulating gas. Both cylinders are heated by electric current. The cylinders are held at 12 m
distance by insulators of epoxy cast resin. The outer cylinder was thermally isolated by glass
wool. The temperatures of the inner cylinder (conductor) and the outer cylinder (enclosure) are
measured at angles of 0° (measure points 1, 2), 90° (measure points 3, 4) and 180° (measure
points 5, 6) by thermo-elements. The far ends of the test set-up did not have any impact on the
measurements and the experimental data was taken after 12 hours since the test started.
  The measured and calculated data are given in Table 3.2.3 for different calculation models.
  The calculated and measured values show only small differences at 0° and 180°, which
are the bottom and top of the gap between the inner conductor and the outer enclosure. The

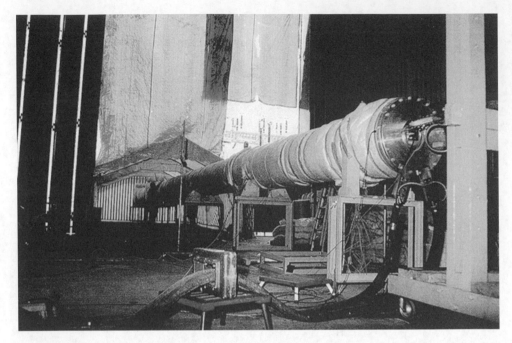

Figure 3.2.9   Experimental set-up for thermal measurements. Reproduced by permission of © Alain Sabot

calculation and measurements at 90° – which is in the middle of the convection – show almost the same values.

Conclusion

The calculations and measurements have shown that, based on a turbulent natural convection in a horizontal concentric ring, the heat transfer between the inner conductor cylinder and the outer enclosure cylinder is very effective at transporting heat away from the conductor. This is the physical basis for the high-power transmission capability of the GIL and its very good overload feature.

The convective heat transfer is active at any load or temperature level of the GIL [96].

Table 3.2.3   Comparison of calculated and experiment results

Angle (°)	Ra $= 2.1 \times 10^9$ Experiment T (K)*	Inner cylinder Equal weight model T (K)	Spalding model T (K)	Van Drist model T (K)
0	325.56	331.39	333.90	333.27
90	325.06	325.06	327.79	327.30
180	325.46	330.67	335.11	335.24

*For numerical calculations absolute temperature (K) are used, while in praxis the temperature is given in °C.

3.2.16.3   Buried GIL

Buried GIL are conducted directly to the soil and the main heat transfer is by heat conductance between soil and enclosure of the GIL. The inside of the gas-insulated GIL uses mainly thermal convection in the gap between conductor and enclosure pipe, as explained in Section 3.2.16.2 and in [52].

The heat transfer in the soil is explained in IEC 60287 [53], with calculation methods. The calculations are based on the physical behaviour that the heat is transmitted to the soil only by conduction.

Finite Element Analysis
To check our analytical model, a finite element calculation was carried out. A 2D model has been chosen since the heat transfer occurs mainly in the radial direction. A calculation domain of 40 m length and 20 m depth is modelled to take account of all the areas influenced by the heat coming from the GIL. The physical parameters are the same as the analytical model. For the mesh, three node elements are chosen and near the GIL the mesh is 10 times denser than in regions far away from the GIL or cable.

Steady-State Results and Discussion
For the calculation, a double system consisting of six single phases has been used, where the distances between the phases are 1.3 m and the depth for the buried GIL is 1.2 m. In Figure 3.2.10 the calculation result is shown on the symmetric line in the centre of the six single phases [52, 78].

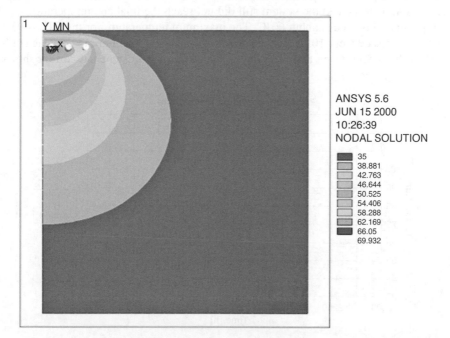

Figure 3.2.10   Temperature distribution of a directly buried GIL by continuous rating: 1700 A, two systems, losses 33.7 W/m. Temprature code: 35° at right side in graphic 66.05° at the inner side of the graphic. All gray levels are in between. Reproduced by permission of © Siemens AG

The calculations are carried out for a constant current load of 1700 A, a constant air temperature of 40°C and a constant soil temperature of 35°C. The thermal resistivity of the soil was also taken as constant, with a value of 1.2 mK/W. The calculation computes the conductor and enclosure temperature and the losses by joule-effect in the heat transfer for the different modes of heat transfer (convection, radiation and conduction).

The middle phase is the warmest one because of the heat input of the neighbouring phases. Maximum temperature is found at the bottom and at the side of the GIL – on the side because of the influence of the other phases, and at the bottom because the thermal resistance from the bottom to the soil surface is greater than from the top to the soil surface. The maximum calculated temperature of the directly buried GIL is shown in Figure 3.2.10, at 69°C.

Overload Calculation for the GIL
Calculations have shown that, even under severe environmental and temperature conditions, the GIL displays overload capability reserves. This means that, even at the constant load of 1700 A, an additional overload can be taken for a period of time. In the following, overload calculations are presented.

Two types of overload have been carried out for two systems of GIL, with axial distances between the phases of 1.3 m and a buried depth of 1.2 m, starting with a steady-state condition of 100%, which is equivalent to a rated current of 1700 A. Calculations have been done for 120, 150 and 200% overload. In Figure 3.2.11 the maximum enclosure temperature of the GIL at 120, 150 and 200% overload is shown [78].

Figure 3.2.11 shows that at a time of about 2400 h, the load was increased by 20, 50 or 200%, which is equivalent to a current rating of 2040, 2550 or 3400 A. The calculation shows that after 2400 h (100 days) the system still did not reach the final end temperature of 69°C. The temperature this time is about 60°C. The maximum temperature for the GIL is 95°C for the enclosure (respectively 105°C for the conductor). This temperature is not reached for an overload of 20 or 50%, even after 133 days. The sloop of the calculated current shows that

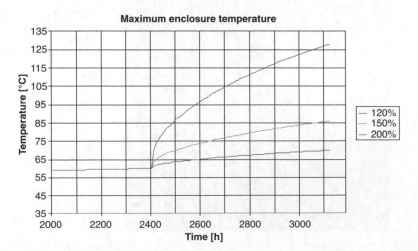

Figure 3.2.11   Maximum enclosure temperature of GIL at 120, 150 and 200% overload. Reproduced by permission of © Siemens AG

the GIL can carry a continuous current of 2500 A, which is an overload of 50% outreaching the maximum temperature. Only with an overload of 200%, which is 3400 A, after 200 h or 8.3 days will the maximum temperature be reached.

Conclusion
In order to describe the thermal behaviour of GIL buried underground in steady state, a computing model has been used. The GIL and surrounding soil are incorporated in the model. It takes into account the three modes of heat transfer conduction in a solid body, convection and radiation. The model used is very comprehensive and simplistic; assumptions have been avoided, especially in so far as the heat transfer occurs mainly in the radial direction and the gas mixture is an ideal gas. The transfer heat across the surface of the GIL is controlled mostly by the convection phenomenon, both in an annulus and from the soil surface to the ambient air. This model was checked by FEM calculations and good agreements were achieved by comparing the analytical and the numerical results.

The temperature at the GIL conductor with the above buried parameters (65°C) remains below the limit fixed for the internal components of the GIL (this limit is fixed by the standard IEC 62271-1 at 105°C for insulators) [82].

The simulation shows that the GIL can be overloaded at 150% with no time limitation without reaching the maximum allowed temperature, and it can be overloaded 200% during 8.3 days to finally reach the maximum allowable temperature of the GIL (105°C).

Among the main factors which influence the heating of a buried system are the following:

*   Thermal resistance in this study of 1.2 mK/W is considered, a value which is used in specifications. The thermal resistivity is supposed constant, but in reality this value increases with time due to the dry-out phenomenon in the vicinity of the buried system, even if a thermally stabilized backfill is used. If a value of 2 mK/W is used in the calculations, the conductor temperature will be increased by c. 30%.
*   The soil surface temperature is supposed constant. If this condition is replaced by a non-isothermal condition (the convection at the interface between the surrounding air and the soil surface), the temperature of the buried system will be increased by c. 10%.

The thermal calculations have shown that with the GIL, a high-power transmission system is available which even under severe climatic and thermal conditions can provide enough safety margin for reliable operation over the lifetime [90].

3.2.16.4   Tunnel-Laid GIL

The thermal design of a GIL laid in a tunnel depends on the transmission losses, the ambient temperature and the length of the tunnel. The heat transfer from the conductor pipe inside the GIL is explained in detail in Section 3.2.16.2. The ambient air around the GIL is then transporting the heat mainly by convection into the tunnel and to the tunnel wall. From the tunnel wall to the surrounding soil, the heat transfer is as explained in Section 3.2.16.3, with the tunnel wall as the thermal conductive heat transfer.

If the heat transfer through the tunnel wall is not sufficient, ventilation of the tunnel is required to transport the heat through ventilation shafts to the outside.

In some cases natural ventilation is sufficient. Natural ventilation occurs automatically, if the ventilation inlet shaft is at a lower altitude than the ventilation outlet shaft (chimney effect).

If this natural ventilation is not sufficient for heat transport, ventilation fans need to be added. Forced ventilation through the tunnel is limited by the wind speed. Low speeds of 5–10 m/s are most effective because of the transfer time for the heat to the moving air stream and to cover the whole tunnel cross-section. The higher the air speed in the tunnel, the smaller the cross-section of moving air and the boundary sections are not cooled efficiently. This means that the forced ventilation has application limits.

If forced ventilation is not sufficient, the last choice is to cool the incoming air. This method is the most costly and needs large cooling units. High investment and maintenance costs are required.

The thermal design of a tunnel-laid GIL is a complex engineering study and it is mandatory for each planning of a transmission project using a tunnel.

3.2.17 Seismic Design

3.2.17.1 General

Besides the mechanical and thermal layout [93], the system behaviour in case of earthquake also needs to be investigated. The GIL is a transmission system over long lengths and its seismic behaviour cannot be simulated by using shaking table and test set-ups. The knowledge of calculation methods to simulate seismic requirements is widely available and proven methods can be used. It is to show that the directly buried GIL withstands the forces and acceleration caused by a typical earthquake scenario. The maximum material stresses of the aluminium conductor using aluminium alloy $AlMgSi_{0.5}$ and of the aluminium enclosure using $AlMg_{4.5}Mn$ must be acceptable in terms of material constant and safety factors. Also, it is necessary to prove a maximum displacement of the conductor towards the enclosure by the given accelerations and frequency modes of this GIL system. The acceleration of the calculated earthquake is taken as 0.5 g in both vertical and horizontal directions.

3.2.17.2 Modelling of the GIL

In the first step, a model analysis has been carried out for the GIL system using the basic elements: conductor, enclosure, conical and support insulators. In a second step, the frequency spectrum has been calculated as a response to a typical seismic stimulation. Finally, the displacement and combination of frequency modes of the GIL system are considered. The spectrum response uses the inherent frequency of the model analysis with the known spectrum to calculate the displacement and the mechanical strength of the GIL system. The spectrum used in this case is the related function between acceleration of the ground and frequency. The assumption in this regulation is that the complete structure of the GIL is stimulated with the same spectrum, which is called a single-point response (SPR).

3.2.17.3 Parameters

The parameters influencing the seismic behaviour of the GIL are the diameter of the enclosure and conductor pipes, the type and distances of the insulators and material constants.

The long length of the investigated GIL section of 120 m makes the relationship to the diameter of 600 mm a long, thin, flexible set-up. The flexibility allows the GIL to follow the

Table 3.2.4   Parameters for the seismic investigations

Conductor	Diameter	220 mm
	Material	$AlMgSi_{0.5}$
Enclosure	Diameter	600 mm
	Material	$AlMg_{4.5}Mn$
Insulators	Support insulator distance	12 m
	Conical insulator distance	120 m
Material parameters	*Aluminium*	
	E-module	$E = 7.1 \times 10^4$ N/mm^2
	Poisson coefficient	$v = 0.33$
	Density	$\rho = 2700$ kg/m^3
	Insulators	
	E-module	$E = 1300$ N/mm^2
	Poisson coefficient	$v = 0.33$
	Density	$\rho = 2700$ kg/m^3

movement of the soil during an earthquake, and damage is unlikely. The flexibility of the GIL, on the other hand, may cause resonances. This may destroy insulators and needs to be avoided.
The parameters chosen for the directly buried GIL for 420 kV are shown in Table 3.2.4.

3.2.17.4   Permitted Stress

The maximum permissible stress of the material with a remaining strain of 0.2% of the plastic strain, including a safety factor of $k = 1.5$, is given in Table 3.2.5. That means:

$$\sigma_a = \frac{Rp_{0.2}}{k} \quad \text{permissible stress} \tag{3.2.1}$$

The conductor is not under pressure, so therefore a safety factor is not necessary.

3.2.17.5   Model of the Calculation

To have a model which is as close as possible to real applications, 10 straight GIL units are welded together to create a 120 m-long GIL section. The post-type insulators are placed every 12 m. The calculations are carried out in three dimensions (x, y, z). The calculations are done by using a modal and spectrum analysis. The advantages of the system symmetry were used,

Table 3.2.5   Maximum permissible strength in the material

	Conductor Type A	Conductor Type B	Enclosure
Permissible material stress (N/mm^2)	$AlMgSi_{0.5}$	$AlMgSi_{0.5}$	$AlMg_{4.5}Mn$
Basic material	160	195	83.3
Welding area	65	95	83.3

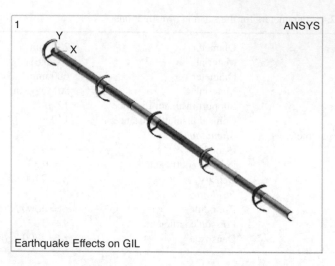

Figure 3.2.12   One-quarter of the 120 m GIL model. Reproduced by permission of © Siemens AG

so only one-quarter of the model needs to be calculated. For the modelling, the geometry was created in Pro/ENGINEER* (a design software) and imported in the form of an IGES file into the finite element program used here to simulate the physical phenomena of the earthquake. A mesh for the calculation is generated by using shell elements. The boundary conditions for the calculations are that the conductor is fixed in the one x direction and free moving in the other, and the conical insulator at the end of the 120 m-long GIL section is a fixed point between conductor and enclosure.

In Figure 3.2.12 the model of the calculation is shown (1/4 of the model). At the beginning of the GIL section the conductor is fixed towards the enclosure. Then, four and a half sections of 12 m follow – where the support insulator is standing inside the enclosure, only half a section in the vertical direction is modelled.

In detail, the model simulates a 12 m-long conductor and enclosure pipes which are welded together to 120 m GIL segments. The post-type insulators are fixed in the x and y direction and have free movement in the positive z direction (see Figure 3.2.13).

3.2.17.6   Analysis Results

At first a modal analysis was done to determine the natural frequencies. The conductor has 120 m length, with a distance between the support insulators of 12 m. This leads to free vibrations of the structure, and then the natural value is in the lower interval, especially in the dynamic resonance range of the spectrum response. The load combinations studied here are reported in Table 3.2.6.

The vertical and horizontal seismic accelerations are supposed to have the same value in order to simulate the worst case. Normally, the vertical acceleration is about 50–75% of the horizontal acceleration. Two arrangements of the insulators supporting the conductor inside the enclosure are investigated. The first one consists of two insulators which are disposed in

---

*Copyright with Pro/ENGINEER.

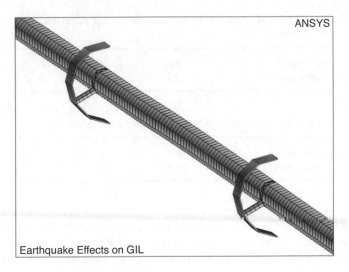

Figure 3.2.13 Model of post-type insulator, enclosure and conductor pipe of 12 m GIL section. Reproduced by permission of © Siemens AG

the lower part with an angle. The second arrangement is with a third insulator in the top of the enclosure. The maximal displacement in the case of the first arrangement (two insulators) is shown in Table 3.2.7.

The maximum displacement in the case* of the second arrangement (three insulators) is shown in Table 3.2.8.

The simulation shows that the conductor oscillates to the top with highest amplitude. The maximum value is in the vertical direction in both cases – vertical and horizontal excitation – because the GIL is free to move in this direction. The maximal amplitude of the GIL with only two insulators as support is 8 mm, while with horizontal excitation the value of the von Mises stress reaches 57 N/mm^2 (this value is 81 N/mm^2 in the case of vertical excitation). This maximum value appears in the contact surface between the conductor and the insulator for the fixed area. The effect of a third support in the top part between the conductor and the enclosure has been simulated. This shows that this case gives a high value of the displacement in the middle part and the stress in the contact surface.

Table 3.2.6 Load factors of spectrum analysis

Load case	Weight	SPR analysis		
		x	y	z
1	1	1	0	0
2	1	0	1	0
3	1	1	1	0

*von Mises stress is an evaluation criteria for numerical calculations.

Table 3.2.7 Maximum displacements in x, y and z direction with two insulators

Direction	ux	uy	zu
Displacement (mm)	0.23	7.6	2.3

Table 3.2.8 Maximum displacement with three insulators

Direction	ux	uy	zu
Displacement (mm)	0.4	19.7	1.9

The combination of the two excitations (vertical and horizontal) is better for the GIL regarding the amplitude of the conductor oscillation ($uy = 6$ mm), but this corresponds to a high value of the von Mises stress (87 N/mm^2), which remains smaller than the stress limit. The theory of modal analysis and the spectrum response doesn't consider the contact elements between the insulator and the enclosure because the spectrum analysis has only a linear behaviour. To solve this problem, the nodes in the contact surface were supposed to have the same displacement in a defined direction. The real displacement with two support insulators is between the calculated value and the value with three insulators. The maximal displacement of the conductor in both cases is between 8 and 19 mm. This value of displacement is the basis

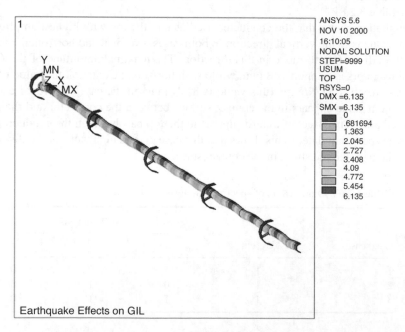

Figure 3.2.14 Displacement in horizontal and vertical directions. Displacement code: Darker grey means higher displacement. Reproduced by permission of © Siemens AG

of the dielectric and mechanical layout of the GIL, so that no electrical or mechanical failure occurs [99].

In Figure 3.2.14 the displacement in the horizontal and vertical directions is shown as a graphical simulation result.

3.2.17.7   Conclusion

The analysis of the seismic requirement of GIL shows that the mechanical forces on the insulators and the pipes are below the allowed stress limits. The oscillating frequencies of an earthquake are much higher than the resonant frequencies of the GIL. This means that in case of an earthquake, the design of the GIL with post-type insulators each 12 m and fixed point conical insulators each 120 m will not be damaged. The mechanical displacement and the material stress are within the given limits. As a consequence, the electrical limits are also fulfilled so that no flashover in the GIL will be expected in case of an earthquake with 0.5 g horizontal and vertical acceleration.

3.3

Product Design

The dimensions of the enclosure and conductor pipes are related to the voltage rating, gas pressure, SF_6 content of the N_2/SF_6 gas mixture and the rated current. GIL is mostly used for the upper high-voltage levels of 420 kV or 550 kV to transmit electric power of up to 3000 MVA per electric three-phase system. In a tunnel, two three-phase systems are usually installed – as shown in Figure 3.3.1.

3.3.1 Technical Data

In Table 3.3.1 the main technical data are given for 420 kV and 550 kV GIL.

At the 420 kV and 550 kV voltage levels, GIL is a single-phase coaxial pipe system of enclosure and conductor pipes made of aluminium alloys. The three-phase GIL, therefore, has three pipes. The high-voltage carrying conductor pipe is held inside the grounded enclosure pipe by insulators. The insulators are post-type and conical-type insulators made of epoxy resin. The insulating gas is 80% N_2 and 20% SF_6 for 420 kV and 60% N_2 and 40% SF_6 for 550 kV. The gas pressure is 0.8 MPa [204, 205].

The enclosure pipe can be made of aluminium sheets which are rolled and longitudinal welded or spiral welded or extruded pipes. Spiral welded pipes are produced with an endless welding machine and cut into the required transport length of 12–18 m. The extruded pipes are produced in the extruder from an aluminium block in the required length and wall thickness.

Inside the enclosure pipe a particle trap is built in along the length to create an area of low electric field to capture free-moving particles. The enclosure pipe is gas-tight and excludes external impacts to the high-voltage part of the GIL.

In Figure 3.3.2 a principle GIL set-up is shown. From left to right: an angle unit, a straight unit, a disconnecting unit, a compensator unit and a straight unit again. These four different GIL units allow us to build any application for power transmission. Including the possibility of elastic bending with a bending radius down to 400 m, the project requirements of a transmission line can be met [206].

Gas-Insulated Transmission Lines (GIL), First Edition. Hermann Koch.
© 2012 John Wiley & Sons, Ltd. Published 2012 by John Wiley & Sons, Ltd.

Figure 3.3.1 Two three-phase GIL in a tunnel. Reproduced by permission of © Alpiq Suisse SA

The principle tasks of the elements are:

straight unit: to cover most of the distances of the transmission line only
                      by bending
angle unit: to make a directional change of any angle

Table 3.3.1 Main technical data

Name	Unit	Variant 1 400 kV GIL	Variant 2 500 kV GIL
Nominal voltage U_r	kV	400	500
Maximum voltage U_m	kV	420	550
Rated current	A	3150	4000
Rated power	MVA	2200	3400
Lightning impulse voltage	kV	1425	1550
Short-circuit current	kA/3 s	63	63
Enclosure pipe diameter	mm	500	500
Conductor pipe diameter	mm	180	180
Gas pressure	MPA	0.7	0.7
Weight per metre	kg	25	30
Wall thickness of conductor pipe	mm	8	10
Wall thickness of enclosure pipe	mm	8	10
Gas mixture	%	20% SF_6 80% N_2	60% SF_6 40% N_2

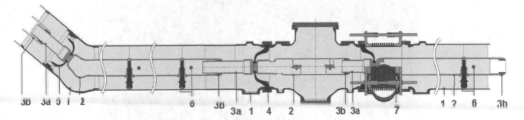

Figure 3.3.2   Principle set-up of the GIL. Reproduced by permission of © Siemens AG

disconnecting unit: to separate into gas compartments
compensator unit: to compensate the thermal expansion of the enclosure when laid in a
                      tunnel or above ground

In the following sections the functions of these GIL units are explained in more detail.

3.3.2   Conductor Pipe

The conductor pipe is an extruded pipe made of electrical aluminium with high electric conductivity, see Figure 3.3.3. Electrical aluminium is chosen to reduce the electric transmission losses by high conductivity. Electrical aluminium has high Al content, usually more than 99.5% to deliver a low electrical resistance for power transmission of low losses. From the the mechanical point of view the electrical aluminium material is very soft on the surface and has low mechanical strength. For that reason electrical aluminium can not be used for enclosure pipes. The length of the extruded conductor pipe depends on the wall thickness and should be as long as possible to reduce the number of joints on-site. The second parameter is the possible transportation length for an assembly on-site. In most cases, lengths of 12–18 m are chosen.

The surface of the conductor joint weld has such a high quality that no additional machine work is used to smooth the surface and to fulfil the high-voltage requirements. For high-voltage reasons the surface of the conductor pipe needs a maximum surface roughness of 10–20 μm and should not have scratches or single pikes, which involves special requirements of the manufacturing process.

3.3.3   Enclosure Pipe

For the larger diameter of the enclosure pipe, different manufacturing processes are available: extruded pipes, spiral welded pipes and longitudinal welded pipes.

Figure 3.3.3   Conductor pipe in the enclosure pipe. Reproduced by permission of © Siemens AG

Extruded pipes are produced by pressing a block of aluminium through a pressuring tool. The large diameter of the enclosure pipe of up to 650 mm requires large pipes of extruder and limits the maximum length of the enclosure pipe. The aluminium alloy needs to have good extruding characteristics, which decrease the electrical conductivity. The pipes are seamless.

Spiral welded pipes are produced by double head arc welding from coil materials; one coil holds several hundred metres of aluminium sheet material. The spiral welding machine produces endless enclosure pipes as long as new coils of aluminium are connected. At the end of the spiral welding machine the enclosure pipes are cut into segment lengths of typically 12–18 m, as required by the project. The pipes are produced with an automated double head welding machine including a double head ultrasonic sensor system for 100% quality control of the spiral weld. In Figure 3.3.4 the outside of a spiral welded pipe is shown.

Longitudinal welded pipes are produced from aluminium alloy plates. The plates have lengths of 3 m and are connected for longer lengths of 12–18 m for the GIL segment. Then, the plate material is formed by rollers to a round shape and a longitudinal weld closes the pipe. The technology of longitudinal welded pipes is seldom used today because there are too many welding requirements. The most used enclosure pipes are spiral welded pipes, which allow high accuracy in roundness and in diameter, which is important for electrical systems. The GIL pipes require accuracies which allow only half the tolerances of standard pipes for no high-voltage application. The tolerance of a 500 mm enclosure pipe is ±10 mm in diameter and roundness. The reason for the large quantities of pipes is that the extruder needs several pipes to be produced before the required tolerances are reached. This is related to the extruder technology.

(a)   (b)

(c)

Figure 3.3.4 Enclosure pipes: A, spiral welded; B, longitudinal welded; C, seamless extruded. Reproduced by permission of © CGIT Systems (a), © Siemens AG (b), © Alstom Grid (c)

3.3.4   Size of Gas Compartment

The size of a gas compartment depends on the availability of gas-handling devices and the maximum allowance of capacitive load of the section when high-voltage tests are carried out. For gas handling, vacuum pumps are needed to produce a vacuum below 2 kPa (20 mbar). This low vacuum requires time for pumping when using gas valves of 25 mm or 40 mm to

Figure 3.3.5   Post-type insulators. Reproduced by permission of © Alpiq Suisse SA

connect the pump. A 1000 m-long gas compartment is seen as a practical length, with vacuum pumping times of 8 to 10 hours.

The same length of about 100 m also fulfils the requirement of the allowed capacitive load of the GIL. The capacitive load gives the amount of energy stored in the GIL. In case of an electric discharge during high-voltage testing, this energy will not damage the insulator surface.

The planning of a GIL transmission line also needs to adapt the gas compartment length to the local situation, where a disconnecting unit which separates the gas compartments can be placed. For this reason the 1000 m length can vary for longer or shorter gas compartments, but not more than 1.5 km length.

3.3.5   Insulators

Since the beginning, different types of insulators have been developed to fulfil the requirements of supporting the conductor and separating the gas compartments. In principle, post-type and conical insulators are used. The post-type insulators can have one, two or three legs (see Figure 3.3.5).

Post-Type Insulators
Post-type insulators may have one, two or three legs depending on their specific task:

- The one-leg types hold the conductor in the centre and enclose the conductor with insulating material. The enclosure side is connected to grounded fixing points.
- Two-leg post-type insulators are plugged into a socket in the conductor pipe and have a sliding mechanism on the enclosure side end. The insulators carry the conductor by its own weight.
- Three-leg insulators are used when the GIL is running vertically and gravity cannot guarantee the fixed position of the conductor in the centre of the enclosure.

The materials used today in GIL are tracking-resistant cast resin mixtures with high mechanical strength. The tracking resistance is important to prevent surface discharges during high-voltage testing.

Conical Insulators

There are two types of conical insulators in use. One forms a gas-tight enclosure of the enclosure pipe at the end, and forms a gas compartment between two gas-tight conical insulators. The other type of conical insulator is non-gas-tight and is used to fix the conductor towards the enclosure, a so-called conductor fix point.

The conus of the conical insulator is different in steepness; typical conus have 20° to 30°, but there are also 45° conus types and at the other extreme a flat plate is used without any conus. The reasons are mechanical strength to hold the conductor even in situations of short-circuit ratings, and to form a gas compartment of up to 1.0 MPa bar overpressure. Depending on the required values, the solution can have different conus types.

In Figure 3.3.6 a conical insulator with a sliding contact is shown. The aluminium enclosure pipe (1) holds, with two fixing rings, the conical insulator (4) inside the enclosure. In the centre the conical insulator holds the aluminium conductor pipe (2) and the sliding contact system with a male contact (2a) and a female contact (2b). The conical insulator has two holes to connect the gas space on the left and right of the conical insulators (4).

If, at this location, the gas spaces left and right are to be separated, then a conical insulator without holes is used to separate the gas compartments. The conical insulator has to take the

1 Enclosure   3a Male sliding contact   4 Conical bushing

2 Conductor   3b Female sliding contact

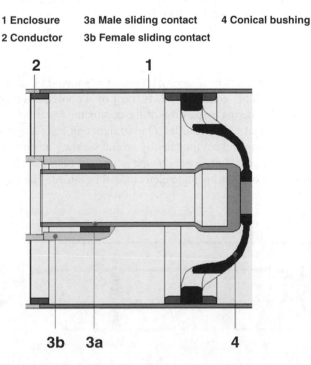

Figure 3.3.6   Conical insulators with sliding contact. Reproduced by permission of © Siemens AG

mechanical forces coming from the conductor in operation (rated current) and also in case of short-circuit currents. For gas-tight conical insulators the maximum pressure (maximum filling pressure on one side and vacuum on the other side) is an additional mechanical design criterion.

The first generation of conical insulators was equipped with ripples to enlarge the creepage distance on the surface. With the improvement of the casting of cast resin insulators and the better quality control of the insulator production process, today these ripples on the surface are not needed anymore.

3.3.6 Sliding Contacts

The current in the conductor will heat up the conductor and the conductor pipe will expand. This thermal expansion is compensated by a sliding contact system. This sliding contact has a plug where a multicontact system can slide on. The advantages of this multicontact sliding system are that the current flow is split by multiple contacts around the conductor pipe and the flexibility of the multicontact allows tolerances for the connection. In many applications this multicontact system has proven its high reliability over the lifetime, which is greater than 50 years.

3.3.7 Modular Design

The GIL has a modular design with only four standard units: straight unit, angle unit, compensator unit and disconnecting unit [38, 100–102].

3.3.7.1 Straight Unit

The straight unit is a single-phase insulated GIL section as shown in Figure 3.3.7.

In the enclosure (1) the inner conductor (2) is fixed by a conical insulator (4) and lies on support insulators (5). The thermal expansion of the conductor towards the enclosure will be adjusted by the sliding contact system (3a, 3b). One straight unit has a length up to 120 m made by single pipe sections jointly welded together by orbital welding machines of GIL segments 12–18 m in length. If a greater directional change is needed than the elastic bending of the straight units can take care of, then an angle element will be added. The angle element covers angles from 4 to 90°.

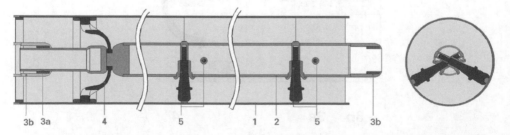

Figure 3.3.7   Single-phase, straight unit of GIL. Reproduced by permission of © Siemens AG

The straight unit only has an insulator positioned inside the enclosure pipe, and the joints are orbital welded. The straight unit can be laid without any outer corrosion protection in a tunnel. When laid directly into the soil, an outer passive corrosion protection is used made of polyethylene or polypropylene coatings. The complete straight unit can be bended when laid with a minimum bending radius of 400 m. When laid in a tunnel, steel structures will hold the straight unit laid on sliding fixing points to allow thermal expansion of the enclosure. When laid in the soil directly, the surrounding soil needs to be free of stones and the straight unit fixed in the ground without longitudinal movement.

3.3.7.2   Angle Unit

Directional changes today use cast aluminium enclosures. Each angle can be designed between 0 and 90°. An angle cut of the pipe is a technical possibility. This is when a pipe is cut at an angle, and then the pipes are turned and welded together again. Another possibility is to use aluminium cast housing with welding flanges at the required angle position. Angle units are used when sharp bends are needed and the elastic bending radius of 400 m is not sufficient. Angle units are more expensive than elastic bending and should be avoided whenever possible.

The angle unit consists of a single-phase enclosure made of cast aluminium alloy. In the enclosure (1) the inner conductor (2) is fixed by a conical insulator (4) and lies on support insulators (5). The thermal expansion of the conductor towards the enclosure will be adjusted by the sliding contact system (3a, 3b). The angle unit is connected to the straight unit by orbital welding, see Figure 3.3.8. The angle element covers angles from 4 to 90°. Under most conditions of the landscape no angle units are needed, because the elastic bending with a bending radius of 400 m is sufficient to follow the contour.

3.3.7.3   Disconnecting Unit

At distances of 1000 to 1500 m, disconnecting units are placed in underground shafts. Disconnecting units are used to separate gas compartments and to connect high-voltage testing equipment for the commissioning of the GIL (see Figure 3.3.9).

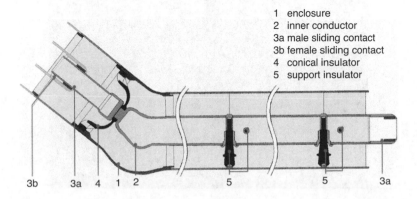

1   enclosure
2   inner conductor
3a male sliding contact
3b female sliding contact
4   conical insulator
5   support insulator

3b   3a   4   1   2   5   5   3a

Figure 3.3.8   Angle unit with straight unit connected. Reproduced by permission of © Siemens AG

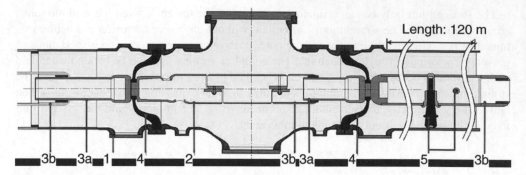

Figure 3.3.9   Disconnecting unit. Reproduced by permission of © Siemens AG

In Figure 3.3.10 the disconnecting unit is shown including a bypass for the gas compartment. This bypass is used to connect the gas compartments before and after the disconnecting unit for operational reasons. In cases of separation of the gas compartments the bypass can be taken out.

3.3.7.4   Compensator Unit

To compensate the thermal expansion of the enclosure pipe, the compensator unit can compensate the length of thermal movement. This compensation is made by bellows, which can take about 300 m to 400 m length of thermal expansion of the enclosure pipe. These compensator units are installed in the tunnel (see Figure 3.3.11). When the GIL is buried directly in

Figure 3.3.10   Bypass of a disconnecting unit. Reproduced by permission of © Alpiq Suisse SA

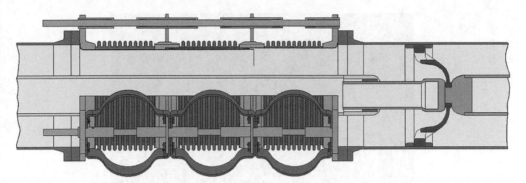

Figure 3.3.11   Compensator unit. Reproduced by permission of © Siemens AG

the ground, these compensator units are not needed because the weight of the soil fixes the enclosure pipe and there will be no movement.

The bellow which takes the movement cannot carry the full electric current, that is why an electric bypass is made by flexible copper bars.

3.3.8   Overhead Line Connection

To connect the GIL to the overhead line, gas to air bushings are needed. These bushings are standard high-voltage bushings. To protect the GIL from transient overvoltages of the overhead line from lightning strikes, surge arresters are connected parallel to the bushing. Bushings and surge arresters can be located underneath the overhead line, as shown in Figure 3.3.12.

3.3.9   Bending Radius

The GIL is a long-distance, underground, high-power transmission technology. This means that it has to follow the landscape. The metallic enclosure and conductor pipes require a laying technique which is similar to oil or gas pipelines. The elastic bending of the material is used. Elastic bending means that there is no plastic deformation, which cannot be allowed because of the high-voltage requirements of the correct roundness of the pipes. Buckling, which is allowed with oil or gas pipelines within limits, cannot be allowed for GIL because the distance between conductor and enclosure is required for electrical insulation. Because of these limitations, the allowed elastic bending is also limited to a smallest radius of 400 m for a 500 mm or 600 mm enclosure diameter.

Besides the elastic bending, the following mechanical stresses also need to be covered by the design:

- gas pressure
- soil load when directly buried
- thermal expansion forces
- thermal stress when directly buried and no expansion is possible
- traffic loads.

Figure 3.3.12 Connection of the GIL to an overhead line. Reproduced by permission of © Alpiq Suisse SA

The bending radius is a basic project planning parameter when the route has to be solved for every metre. The total project costs are related to the number of required GIL angle units when the directional change cannot be solved by elastic bending below a radius of 400 m.

In Figure 3.3.13 the elastic bending is shown when a GIL is laid in a tunnel. The bending radius in this case is 700 m.

3.3.10 Joint Technology for Conductor and Enclosure

The connection technologies available for GIL are:

- flanges with bolts and O-ring sealing
- welded
- clamps.

The choice of the right technology depends on the requirements. Flanges with bolts and a single O-ring sealing have been used in GIS for more than 40 years and show high quality and

Figure 3.3.13   Bending radius of 700 m in a tunnel. Reproduced by permission of © Alpiq Suisse SA

reliability. The disadvantage when long distances have to be covered is the intensive manual handling on-site when the GIL is assembled. The flanged technology is used for the enclosure joints to form gas compartments under pressure.

For long distances, an automated orbital welding system offers less manual handling on-site and a better repetitive quality including an ultrasonic quality insurance system. Welding technology can be used for gas-tight enclosure joints and for electrical connections of the conductor pipes.

The clamp technology can be used for electrical connection of the conductors. Clamp technology is also well known and used with high reliability. Similar to the flange technology with bolts and a single O-ring sealing, the clamps are also connected manually. The clamp technology can be designed to standardize the connecting work steps and minimize the manual work on-site.

A new technology which is very robust for on-site application is friction steer welding. This welding process uses a stir to generate heat, to melt the aluminium and to make a holomorphic connection. The advantages of this technology are: higher speed and almost no impact from humidity and moving air.

3.3.10.1   Flanged Joints

The jointing technology used in gas-insulated switchgear is the flange with bolts connected and sealed with an O-ring. This technology is well proven and has been applied in millions of joints which all show good experiences.

The flange technology is also used in many GIL applications, mainly in those cases where the system length is not high and may also be spread over a substation (see Figure 3.3.14).

Figure 3.3.14   Flanged joint. Reproduced by permission of © Alstom Grid

3.3.10.2   Arc-Welded Joints

Arc welding is a well-known technology from pipelines and also in power plants. For use in substations and mainly when longer distances need to be covered, the application of aluminium pipes is necessary. The welding of aluminium pipes is different from steel, and the welding parameters have to be adapted. In orbital welding technology the welding process is highly automated. Only a few parameters are directly influenced during the welding process on-site by the welding engineer. Orbital welding produces several layers to guarantee the gas-tightness of the GIL [76, 79, 107].

Application
The orbital welding of the enclosure pipe needs to guarantee gas-tightness and mechanical strength. The welding seams need to be prepared a short time before the welding is carried out to prevent the aluminium surface from corroding. See Figure 3.3.15.

The prepared pipes are held with clamps to fit the welding seams together. The pipes are concentrically fixed and centralized by a lost welding ring. The chosen welding process is tungsten ignition gas (TIG) welding, which shows good welding results also on-site, when ambient conditions may change with the weather.

Figure 3.3.15   Welding preparation. Reproduced by permission of © Siemens AG

All the quality tests showed that the TIG welding can perform orbital welding which fulfils the requirements of gas-tightness and mechanical strength. The welding process requires a welding tent to protect the welding process against air movement.

The orbital welding has been automated as far as possible to provide a failure-free, standardized and repetitive weld for the enclosure and the conductor pipe. The arc welding is explained by 95 different parameters which need to be optimized for each welding process of a given pipe material, wall thickness and diameter: 91 parameters are fixed in the control computer of the orbital welding machine; 4 parameters are left for manual control of the welding technician (speed of the welding head, speed of the welding wire, current of the welding arc and impulse time of the welding current). See Figure 3.3.16.

To change the welding process from enclosure to conductor pipe welding only the guiding ring for the weld head and the software of the control computer need to be changed (see Figure 3.3.17).

3.3.10.3   Friction Stir-Welded Joints

Friction stir welding (FSW) was developed and patented for aluminium alloys in 1991. First applied to plates and straight welds, it is now also available for pipes. FSW uses a stir weld head in a cage around the pipe and is powered by a hydraulic aggregate. Pipes of 500 mm or 600 mm diameter of $AlMg_3$ or $AlMg_{4.5}$ can be used.

The main advantage of FSW is the higher welding speed, because the complete wall thickness of the pipe can be welded at once and the FSW welding process is independent of the ambient air condition. Air movement does not affect the welding process as it does with arc welding. Standard EN 288 gives the quality requirements. The time saving of FSW welding over arc welding is about 40%.

FSW does not melt the aluminium to connect the pipes; it works below the fluidity temperature. The aluminium material is made soft and connects the material before it is fluid.

Figure 3.3.16 Orbital welding process controlled by welding technician. Reproduced by permission of © Siemens AG

FSW uses a cylindrical tool of polished steel. When rotating into the aluminium it makes the material soft by generating friction heat and mixing the softened material. The rotating steel tool is pushed forward along the cylindrical shape of the pipe when the material is soft. Right after the rotating tool the material cools down and creates a solid connection of the two pipes. In Figure 3.3.18 the principle is shown [104, 105].

FSW uses mechanical forces to create the heat from the friction of the rotating steel tool to make the material soft and to connect the pipes. Pipes of 500 mm and 600 mm diameter with a wall thickness of 10 mm have been successfully connected. To bring the mechanical forces into the material, clamping tools are needed around the pipe to avoid determinations of the pipes. One ring is used to lead the FSW head around the pipe and a second ring is used to hold the pipe in its round shape. The required forces for softening the aluminium come from the forward speed and the rotating speed of the FSW head.

The FSW process is automated with electronic control of the hydraulic drive. The ring holding the FSW head and an integrated camera can be split and laid around the pipe. In Figure 3.3.19 the FSW orbital welding machine is shown. An integrated camera controls the start of the weld when the FSW weld head enters the material.

The aluminium pipes are delivered in diameter tolerances of 1%, which is too much for the FSW process. Therefore, the welding ends of the pipes need to be brought to better tolerances. This is done by a hydraulic tool which forms the ends to diameters of the required tolerance by expansion (see Figure 3.3.20). With hydraulic pressure of 50 MPa the exact diameter of the pipe is produced.

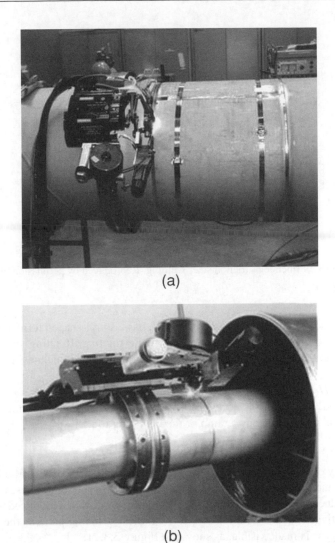

(a)

(b)

Figure 3.3.17 Orbital welding procedure: (a) enclosure pipe; (b) conductor pipe. Reproduced by permission of © Siemens AG

The plastic formed end of the pipe is expanded by 30 mm for a 600 mm diameter. The diameter tolerances after the expansion are less than ±0.4 mm.

Lost Weld Ring
The mechanical forces from the FSW head cannot be taken by the aluminium pipe. Therefore, a lost welding ring is brought inside the aluminium pipe to prevent the pipe from deformation. This steel ring is called lost because it will remain inside after the weld is completed. The steel ring is shrunk into the aluminium pipe by heating the aluminium. When the steel ring is inside, the aluminium cools down and makes a solid connection to the steel ring.

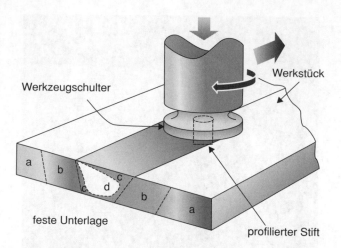

Figure 3.3.18   Principle of friction steer welding. Reproduced by permission of © Areva

FSW Tool Head
The FSW tool head uses a polished steel pin which has a spiral-shaped form and is positioned with 2–4° angle to create a forward movement while rotating. Rotating speed and forward movement need to be adapted to the aluminium material and wall thickness of the pipe.

Closing the End Hole
To close the weld at the end it is necessary to move the FSW head pin while rotating out of the material. This is controlled by a ramp of the clamping tool.

Testing the Weld
The weld quality can be checked by X-ray or ultrasonic measurements. These methods show inner porosities and give an indication of the correct quality of the weld. These tests are carried out on each produced weld. For the proof of the welding process, the produced weld is cut into slices and investigated under the microscope. With these grinding patterns the correctness of the welding process is made visible, as shown in Figure 3.3.21.

Mechanical strength is tested by bending tests according to standards. These bending tests show possible weaknesses of the material, and the welding process is only qualified when all of these tests have been passed.

The parameters to be adjusted to reach an acceptable quality of the weld on aluminium pipes are the speed of the rotation and the forward movement. Values of 20–120 mm/min and rotating speed of 300–500 turns per minute are the typical. Also, the length and size of the rotation pin are of importance for the weld quality. Depending on the wall thickness, the rotating pin needs to be shorter than the wall thickness.

3.3.10.4   Ultrasonic Test

Each weld is quality controlled by an automated ultrasonic test. The ultrasonic test head is fixed by a guide ring at the weld and controls the complete weld by moving once around the

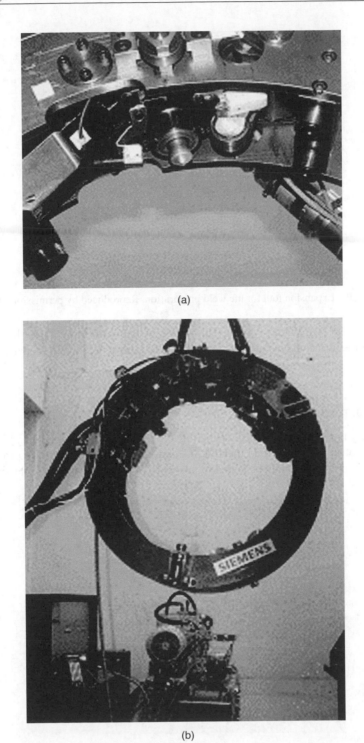

(a)

(b)

Figure 3.3.19 FSW ring including welding head and control camera: (a) welding head; (b) welding
ring. Reproduced by permission of © Areva

Figure 3.3.20 Expansion tool for the weld preparation. Reproduced by permission of © Areva

pipe's orbital weld (see Figure 3.3.22). This 100% weld control protocol is then stored with each weld. If the ultrasonic test detects a welding failure, the area will be repaired by manual welding.

3.3.11   Corrosion Protection

For the directly buried GIL in the soil, a coating against corrosion is absolutely necessary and similar to the other directly buried metallic installation. An active corrosion protection, also called cathodic protection, may be added to take care of possible failures of the coating as an

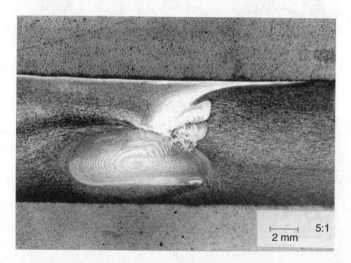

Figure 3.3.21   Grinding pattern of an FSW. Reproduced by permission of © Areva

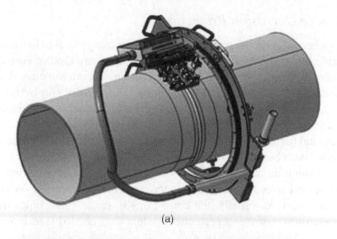

(a)

(b)

Figure 3.3.22 Ultrasonic test of the weld: (a) principle; (b) equipment. Reproduced by permission of © Areva

additional quality insurance system to the passive corrosion protection. Failures in the passive corrosion protection may occur over the lifetime of the system by outer damage through other earth works or by cracks or voids in the material. Then, the active corrosion protection system protects such cracks and voids in the passive corrosion system against corrosion. This section proposes some solutions able to protect the GIL against corrosion by using both passive and active corrosion protection [106–109].

3.3.11.1 Passive Corrosion Protection

For applications where aluminium pipes are used in air above ground or in a tunnel, aluminium generates an oxide layer which protects the enclosure from any kind of corrosion. The oxide layer of an aluminium pipe is very thin – only a fraction of a micrometre. But it is very hard and very resistive against a gaseous environment like the atmosphere. In most cases it is not necessary, even in outdoor applications, to protect the aluminium pipes against corrosion with, for example, coatings [106].

Going underground for directly buried systems, the corrosion situation changes dramatically. Continuous contact with humidity, water and all kinds of chemical elements can destroy the oxide layer of the aluminium pipe and then cause corrosion. Different from steel pipes, the corrosion is not spread over an area but is very concentrated, centred on one place where the oxide layer is destroyed. Finally, the corrosion will cause small holes in the enclosure. To avoid such corrosion, corrosion protection systems are needed (see Figure 3.3.23). Two

Figure 3.3.23 On-site corrosion protection of the welding area – shrinking method. Reproduced by permission of © Siemens AG

basic methods are used today: passive corrosion protection and active corrosion protection. The passive corrosion system is an added layer of non-corrosive materials, for example polyethylene (PE) or polypropylene (PP), while the active protection system uses a voltage protection level of the protective aluminium enclosure towards an electrode.

In this case a corrosion protection system based on the shrinking method is used. Other methods use granulates or tapes and are standard knowledge in the pipeline industry.

3.3.11.2   Active Corrosion Protection

Using active corrosion protection the induced current generates a voltage potential of the metallic enclosure towards the soil. If this voltage level is at a potential around 1 V towards a loss electrode, then the loss electrode will always corrode and not the aluminium enclosure. The active corrosion protection system is an additive to the passive corrosion protection system. The active corrosion protection system takes care of possible failures of the coating as an additional quality insurance system to the passive corrosion protection. Failures in the passive corrosion protection may occur over the lifetime of the system by outer damage through other earth works or by cracks or voids in the material. Then, the active corrosion protection system protects such cracks and voids in the passive corrosion system against corrosion.

Experience world-wide with buried pipe systems installed directly underground shows that over the decades of underground laying systems some cracks or voids in the corrosion protection may occur, which adds up in an increase of the induced current of the active protection system. So, the positive effect of the active corrosion protection system is that not each failure needs to be repaired immediately, and a guaranteed lifetime of the passive corrosion protection system of 30 years can be extended by many more years.

The active corrosion protection system, also called cathodic corrosion protection, uses an induced current to adjust the protective voltage of approximately 1 V against the loss electrode. To reach this 1 V protective voltage an induced current of approximately 100 μA is needed. The induced current is related to the total of the surface to be protected and will increase with the length of the system and the number of failures. In practice, several kilometres can be protected with only one DC voltage source because the current is low.

Directly buried GIL have been in operation for more than 20 years, in the United States inside transmission substations.

In the IPH test field in Berlin, Siemens installed and a consortium of three German utilities tested an approximately 100 m-long directly buried GIL. The prototype was installed to carry out a long-duration test to simulate a 50 years' lifetime of the system. With this long-duration test the passive and active corrosion systems were also tested. In Figure 3.3.24 the test set-up is shown. At the top of the figure the site view of the installation is shown. On the left side through a shaft the GIL is connected to the high-voltage and high-current sources. The GIL is then bent with a 400 m bending radius to the side and down until an angle element makes a directional change. See side view, upper half of Figure 3.3.25 and top view, bottom of Figure 3.3.24. The test set-up ends at the other end in another shaft [108].

Along the test installation are placed the temperature sensors (sensors T01 to T14), the sensors for the internal gas pressure (sensors P01 to P14), the sensors for the displacement (sensors B1 to B6) and finally the connection of the cathodic corrosion protection in shafts at the end. Long-duration tests on the buried GIL have been carried out under real on-site

Figure 3.3.24 Test set-up for 100 m-long directly buried GIL. Reproduced by permission of ©
Siemens AG

conditions. The corrosion-protected GIL has been laid on a sand bed and was backfilled with
the soil after the laying process was completed.

3.3.12   On-Site Assembly Work

The pre-assembly of the GIL segments is not done in the factory because of the difficulties
of transporting the GIL segments on-site. Cost and risk of damages are high. That is why,
for large-scale projects when long distances need to be built, the GIL segments are produced
close to the site. The single elements are therefore delivered direct to the construction site
and assembled in transportable assembly tents. The conditions on-site need to be the same
for high-voltage clean assembly as in the factory. The assembly tent is strongly controlled in

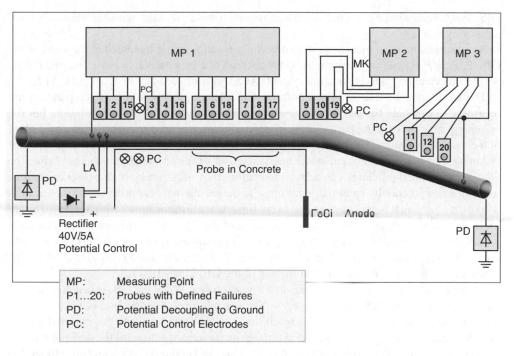

Figure 3.3.25   Corrosion protection monitoring includes temperature, pressure and displacement sensors. Reproduced by permission of © Siemens AG

terms of clean conditions, and the high-voltage compartments are always protected against dust and humidity. The enclosure pipes are kept closed for protection against particles and dust and only opened when needed for jointing [103].

With this laying method a high productivity on-site can be reached. This is necessary when longer lengths of 10 km or more need to be assembled.

To reach the clean requirements necessary for high-voltage equipment, special assembly areas are defined inside the assembly tent. Areas where metallic particles are produced are separated from the high-voltage assembly areas. Dust and humidity are controlled by air conditioning where necessary. The single steps in the assembly are precisely adapted to the need for cleanliness.

The position of the assembly tent on-site depends on the on-site conditions. The tent can be placed in the tunnel, the shaft of the tunnel, or in or beside the trench when directly buried. The key issue for assembly on-site is cleanliness. Cleanliness needs to be ensured at each step of the assembly process. The solution for conditions on-site can vary significantly, and in any case needs to be planned and engineered before work starts.

3.3.13   Monitoring

For monitoring and control of the GIL, secondary equipment is installed for measurement of the gas pressure and the gas temperature. These are the same elements which are used in

GIS. For commissioning, partial discharge measurements are used with the sensitive UHF measuring method.

An electrical measurement system to detect arc location is implemented at the ends of the GIL. Electrical signals are measured, and the position of a very unlikely case of internal failure will be calculated by the arc location system (ALS) with an accuracy of 10 m [111, 112].

The monitoring of GIL is very similar to GIS, where gas pressure and temperature are measured to calculate the gas density as the requirement for high-voltage insulation. For the commissioning process the partial discharge measurement is used to identify internal failure of the insulation system. The partial discharge measurement system can identify if an insulator is damaged, if a conductor or enclosure has damage on its surface, or if particles are inside the GIL. If indicated, the failures can be reported before they occur under high-power conditions in service and potentially cause large failures. To detect internal arcs precisely at the location where they happen, it is necessary to have a continuous arc location monitoring system. The arc location system is triggered by the travelling wave of the GIL and calculates the fault location within 10 m accuracy. This is an important item of information for a quick report process. The information is automatically indicated in the control centre. The measuring sensor is a VHF antenna, which is placed in a flange at the end of the GIL [76, 101, 110].

Moisture in the insulating gas is measured by gas probes. When the GIL is installed and the gas compartments are filled with SF_6 or N_2/SF_6 gas mixtures, the dew point of the moisture measurement is set below $-20°C$. To reach this low moisture level the gas compartment is prefilled with dry nitrogen and circulated through an gas drying unit until the dew point of the gas is below $-20°C$. Gas probes are taken after filling, and at intervals of some days to see if – from the first filling with dry gas – the moisture content has increased. Increasing moisture would indicate that there is humidity stored in the insulating material, which comes out of the material slowly. The long-time experiences with gas-insulated systems, over more than 40 years, now show that once the internal moisture is stable the GIL will be okay for its lifetime. The lifetime today is 50 years, but it is expected by experience that the real lifetime will be longer [47, 103].

In Figure 3.3.26 an overview is given of the components of a GIL monitoring system.

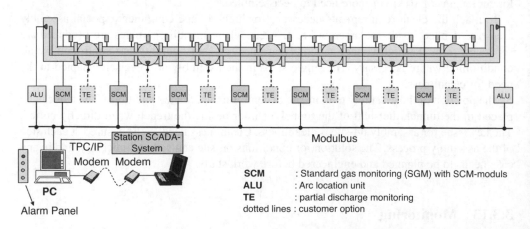

Figure 3.3.26 Components of GIL monitoring system. Reproduced by permission of © Siemens AG

Figure 3.3.27   Gas density monitoring device. In praxis the indication is: green – okay (right side of scale) red – alarm (left side of scale). Reproduced by permission of © Siemens AG

3.3.13.1   Gas Density Monitoring

The gas density is the physical value guaranteeing electrical insulation. Gas density can be calculated from the gas pressure and the temperature. When the gas density is declining because of a leakage, the monitoring will first give a warning and when the gas leakage is continuous, the monitoring system will give the order to switch off the GIL. There are two values specified for the GIL. The first value is, for example, 10 kPa bar below the filling pressure; the second value is, for example, 20 kPa below the filling pressure. The filling pressure is typical 0.8 MPa.

To simplify the gas density measurement, in most cases the density value is not shown as an absolute value. The indication that the system is okay is shown in the colour green, and at the density for a warning of gas losses the indication is at the transition between green and red. Finally, when the line is shut down the indication will be fully in the red (see Figure 3.3.27).

Figure 3.3.28   UHF antenna connector at the flanged enclosure plate. Reproduced by permission of ©
Alpiq Suisse SA

3.3.13.2   Partial Discharge Monitoring

Partial discharge (PD) monitoring is a method to control and measure the dielectric integrity
of the high-voltage system. There are PD systems of different physical nature available. The
classic method uses high-frequency impedance in the ground connection of the electric system
to measure the signals caused by internal electrical partial discharges. The partial discharge
currents are typically of some nano-amperes only, with measured capacitive displace loads of
pico-coulombs (pC), as a apparent charge.

A second method is based on acoustic waves, which are caused by partial discharges in the
gas or in the insulator. The acoustic waves are measured with sensors at the metallic enclosure.

A very sensitive method of measurement is the UHF sensor method, where an antenna
inside the high-voltage gas compartment measures directly the radio frequency limited by the
partial discharge in the gas or inside the insulator. This method is seen as the most sensible
and can easily be adapted to the GIL with its long coaxial gas compartments.

From the outside only the antenna connector can be seen. The UHF antenna is fixed at the
inside of the disconnecting unit flanged enclosure plate (see Figure 3.3.28).

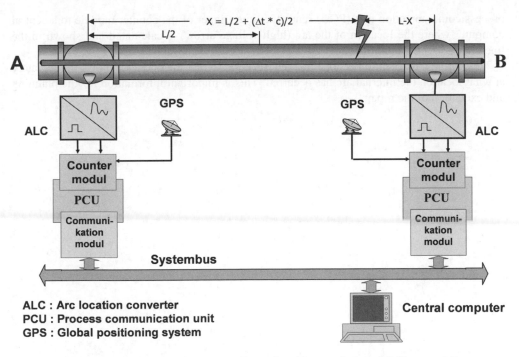

Figure 3.3.29 Arc location system. Reproduced by permission of © Siemens AG

The sensitive measuring technology of UHF partial discharge monitoring is an important quality tool to prove the high-voltage quality after assembly of the GIL on-site. It is not needed for permanent PD measurements [70, 112, 214, 219].

3.3.13.3 Arc Location System

The internal arc fault location uses the measurement of the time difference of the surge impedance travel wave from the location of the arc to the end of the GIL. The physical length and electrical impulse propagation times are known, so that from the different arrival times of the travelling wave at the ends of the GIL the exact location can be calculated. The time accuracy is about 80 ns and the accuracy in the location is about ±25 m. This does not depend on the length of the GIL line. The time synchronization can be made by an onboard clock or the GPS time signal. The result is indicated in the control centre.

In Figure 3.3.29 the ALS is shown in principle. The arc location uses the travelling time of the electrical impulse inside the GIL. The GIL length between point A and B is known and the travelling speed of an electrical impulse, for example an internal arc (high voltage arrow), is also known as close to the speed of light.

Depending on the arc location, the distance to the end points A and B is fixed. The sensors at both ends, called arc location converts (ALC), transfer the travelling impulse in the GIL into a trigger signal which triggers the GPS time synchronizer in the time counter module. A process communication unit (PCU) sends the time trigger of the counter module to a central computer where the location of the arc (high voltage arrow) is calculated and shown in the line diagram of the control centre.

This whole identification is done in a few seconds so that the operator knows immediately at what location the internal arc has occurred. This is important information for the planning and execution of the repair.

3.4

Quality Control and Diagnostic Tools

Quality control and diagnostic tools are used for the correct on-site assembly of the GIL and to avoid or correct assembly failures. The measurements of gas mixtures using SF_6 and N_2, as explained in Section 3.1, have shown that the sensitivity of gas mixtures towards failures (like inhomogeneous fields around scratches or edges) is similar to that of pure SF_6. This means that the experience gained with pure SF_6 in dealing with failures in the dielectric system can also be used for N_2/SF_6 gas mixtures.

The usual surface roughness of metal surfaces in N_2/SF_6 gas mixtures, in comparison with pure SF_6, is the same or shows a lesser reduction in dielectric strength [59], whereas larger metal protrusions or mobile particles are more critical [75]. Mobile particles inside a GIL have to be limited in seize and in number. To reach this the assembly process is following clean assembly rules and the GIL is equipped with particle traps.

The statistical scatter of the breakdown voltage was found to be similar for all mixtures and for pure SF_6 of equal dielectric strength [61]. Therefore, the approved conventional test procedures for GIS can be applied to confirm the required withstand levels [59]. For insulation coordination of extended transmission lines, the Weibull distribution has to be applied [91].

By means of careful assembly work and efficient quality control, defects are seldom – indeed practically ruled out. Mobile particles are the most frequently remaining type of defect. They are usually eliminated finally by conditioning procedures during power/frequency high-voltage testing [76]. In a conventional GIS with a complex insulation system they are moved by a stepwise increased AC field stress into low field regions, which act as natural particle traps. In a plain insulation system of a GIL with enclosure pipes and conductor pipes only there is no natural particle trap. Therefore, the GIL has been equipped with artificial particle traps all along the GIL extension, which proved to be very efficient.

Modern diagnostics are applied for the detection, localization and identification of defects. The UHF method proved to be most efficient [76]. Its application is restricted by signal attenuation and the correspondingly limited measuring range of installed sensors. In GIS, this attenuation is mainly caused by the usually installed conical spacers. The maximum distance between sensors should therefore normally not exceed 20 m. The predominantly used slim

Gas-Insulated Transmission Lines (GIL), First Edition. Hermann Koch.
© 2012 John Wiley & Sons, Ltd. Published 2012 by John Wiley & Sons, Ltd.

support insulators in the Siemens GIL design cause almost no signal attenuation, since at the insulator location almost the whole cross-section is gaseous and open for electromagnetic signal transfer. Therefore, in the GIL an efficient UHF PD measurement can be carried out even with distances between sensors of several hundred metres. This enables the use of the UHF method in GIL, as has been successfully performed for the first time on-site at the PALEXPO project in Geneva in 2001 [113].

Moisture penetration by diffusion through the enclosure and from the bulk of the insulators into the gas is, in GIL, much less than in conventional GIS, due to the excellent gas-tightness of the welded joints and the low amount of solid insulating material which may contain moisture. The insulation quality of the insulator surfaces can therefore reliably be preserved by conventional measures to avoid dewy surfaces of reduced dielectric strength.

Altogether, it can be expected that the GIL will give the same or even better long-term performance than the GIS, which demonstrates a long service lifetime with no critical ageing even after 30 years of operation [35], because in the GIL almost the same materials are used, the amount of solid insulating material and SF_6 is considerably reduced and the approved quality control by way of tests and modern diagnostics can be applied [61].

3.4.1   Quality of Parts

Each single part of the GIL has to pass a quality control which is adapted to the high-voltage requirements and the related functions.

Insulators are produced in a mostly automated production process, which delivers a constant quality. At the end of the production process the final quality control on each insulator is done by partial discharge measurements, and the required level has to be below 5 pC, as required by the IEC standard for GIL [54]. This required partial discharge level of 5 pC proves that the insulators are free from internal failures that could be harmful during their lifetime.

Gas-tight insulators need to withstand the filling pressure of about 0.8 MPa. Each single gas-tight insulator is pressure tested according to the relevant pressure vessel standard [83–88]. This quality test is a mandatory test because it is safety relevant.

The conductor pipe, the sliding contact system and any metallic part needs to be free of surface failures. This test of quality is made via view control by experts with high-voltage knowledge and experience. The high-voltage knowledge is a basis for failure-free assembly.

3.4.2   Quality of Processes

The different processes for welding, assembly on-site and laying are part of the quality control process on-site. The ultrasonic test procedure of the orbital welding is explained in Section 3.3.10. When the GIL is assembled on-site by orbital welding, each weld is 100% quality controlled by ultrasonic measuring equipment. The complete data is stored, and part of the project documentation. Only failure-free welds get approved. If failures like voids are detected, a manual hand weld will correct this failure.

The assembly process on-site is quality checked at each step of the assembly. Prefabricating the GIL elements in the site close assembly tent, transportation from prefabrication to the laying location, and the laying process itself. The principle of the high-voltage conform assembly process is to keep the high-voltage gas compartment clean and always closed to avoid particles and moisture entering the gas compartment.

In this book it would be too detailed to explain all the single steps of quality control checks, and also this is part of the knowledge of the manufacturers. All of the quality checks are documented with the project documentation, so that at the end – for each GIL element – the quality control results can be seen.

The final quality control check of the laid GIL is the on-site high-voltage test, where 80% of the design test voltages are applied to the GIL. The advantage of GIL is that before starting operation, each section has passed a high-voltage test. Also, for this reason, the maximum length of a section (gas compartment) is limited to 1.5 km.

The on-site testing and long-duration tests of GIL are explained in Section 3.8.

3.4.3   Partial Discharge Detection

Partial discharge detection is an important quality control before the GIL is energized to prove the correctness of the assembly process. To identify failures of the insulation, the PD detection system offers a high sensitivity to identify partial discharges over distances inside the GIL of up to 500 m. The UHF antenna, which operates in the frequency range of several hundred megahertz to gigahertz, can detect partial discharges over such long distances. The reason for this high sensitivity is because the coaxial and cylindrical GIL is an excellent high-frequency conductor with low impedance. In Figure 3.4.1 the result of PD measurement with a spike on the conductor is shown.

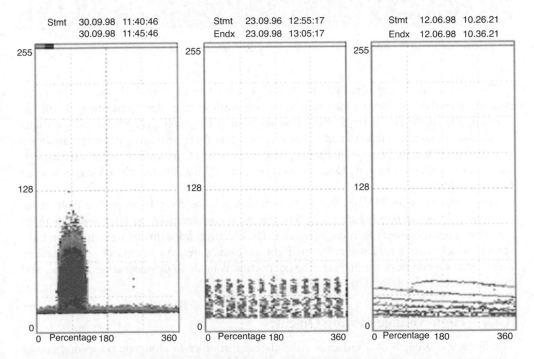

Figure 3.4.1   PD measurements. Left: spike on the conductor, middle: at 70 m distance, right: noise. Reproduced by permission of © Siemens AG

Figure 3.4.2   UHF sensor for PD measurements. Reproduced by permission of © Siemens AG

The UHF method is also usable in a N_2/SF_6 gas mixture [4]. The UHF partial discharge measuring method can identify internal free-moving particles (e.g., dust) when the high-voltage is first applied to the GIL in steps. With each step of increasing voltage, a PD intensity is found which disappears after some minutes. The reason is that these particles will be captured by a so-called particle trap. The commissioning process uses the PD measurement for a secure start for the high-voltage operation. Once in operation, the PD measurement is no longer needed [112, 214, 219].

The UHF sensor for PD measurements is mounted inside the GIL through a plate at the end of a GIL section (see Figure 3.4.2). The electrical connection of the UHF sensor is made through a pressure-proof electrical connector. The electrical adjustments for setting the UHF sensor, which acts as an antenna, to the right measuring frequency band of some hundred megahertz is made from the outside. The connection is made to the measuring amplifier and evaluation computer system by coaxial wire.

3.4.4   High-Voltage Testing On-Site

The dielectric strength of a GIL has to be checked in order to eliminate accidental causes (damage during handling, transportation, storage and erection) which might give rise to an

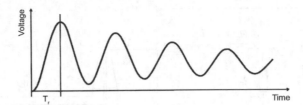

Figure 3.4.3   Oscillating impulse voltage for on-site testing: t, time; T_f, front duration; V, voltage

internal fault in service. Because of their different purposes, these tests will not replace the type tests or routine tests in the factory. They are supplementary to the dielectric routine tests with the aim of checking the dielectric integrity of the completed installation and detecting irregularities, as mentioned above [115].

For on-site tests, GIL can be divided into sections, taking into account the availability of test equipment.

For the choice of an appropriate voltage waveform, IEC 60060-1 [116] should be taken into consideration; however, similar waveforms are also permissible. AC is preferred, with PD monitoring during application of the test voltage. Conventional PD measurement in accordance with IEC 60270 [117] will not usually be appropriate. Other methods, such as UHF methods, should be considered.

In many cases, additional impulse testing may be desirable. The following gives some aspects on the choice of the waveform. Figure 3.4.3 shows a typical impulse voltage which has been used for on-site testing of GIS and will now be applied for GIL.

The test and measuring facility for this oscillating impulse voltage is relatively lightweight and easy to handle. It consists of a conventional impulse generator, a long single-layer coil, and both damping and discharge resistors. The capacitance of the tested device determines the series resonant frequency in combination with the oil inductance.

From dielectric studies on defects in GIS it is well known that lightning impulses are best suited for finding sharp protrusions in the electrodes and particles on insulator surfaces. For long test objects, however, the front duration must be long enough that the influence of travelling wave phenomena is kept low enough and the complete test object is subjected to the same voltage amplitude.

The following example (Figure 3.4.4) shows the mentioned effects [47, 76, 114, 146]. The length of the tested GIL section is assumed to be 1 km. The front duration is mainly determined by the coil inductance. A first calculation is aimed at a front duration slightly above 10 µs. Date for the calculation is as follows:

- Impulse capacitance C_s 1.2 µF charged to 1 p.u.
- Coil inductance L_s 0.2 mH or 1.7 mH
- Damping resistance R_s 20 Ω
- Discharge resistance R_D 100 Ω
- Surge impedance of GIL   55.4 Ω

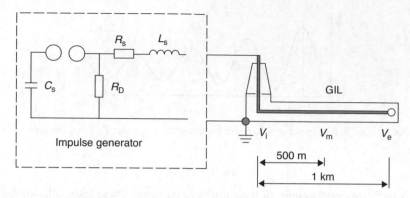

Figure 3.4.4 Circuit for oscillating impulse testing

Figure 3.4.5 shows the result of the calculation of the voltages at the GIL input, in the middle of the GIL and at the open end of the GIL test section. It is evident that the differences in the voltage amplitudes at different locations cannot be tolerated for testing purposes.

Therefore, a coil inductance of 1.7 mH was used for another calculation. All other parameters remained the same as before. Figure 3.4.6 shows the resulting voltages V_i, V_m and V_e; the variations in amplitude are small enough for on-site testing.

The on-site high-voltage test equipment is shown in Figure 3.4.7 when delivered on-site by truck. The size and volume of such test equipment place limitations on the maximum length of the GIL sections. With a typical capacitance of 55 μF per kilometre GIL single-phase pipe length, the maximum length for test equipment available today is about 1.5 km.

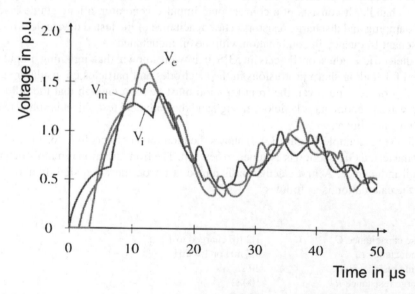

Figure 3.4.5 Voltages at three different GIL locations with $L_s = 0.2$ mH: V_i, voltage at the GIL input; V_m, voltage in the middle of the GIL test section; V_e, voltage at the end of the GIL test section

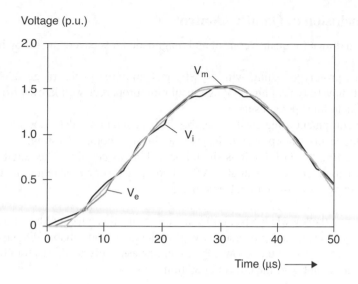

Figure 3.4.6 Voltages at three different GIL locations with $L_s = 1.7$ mH: V_i, V_m and V_e show small differences

Figure 3.4.7 On-site high-voltage test equipment. Reproduced by permission of © Siemens AG

3.4.5   Conclusion of Quality Control

The conclusion drawn for quality control and diagnostic tools covers the following aspects:

- 420 kV GIL prototype testing with N_2/SF_6 gas mixtures in the range of 10% to 20% SF_6 content showed typical high-voltage insulation properties well known from 100% SF_6 insulation, including epoxy insulators.
- Monitoring equipment using easily available instrumentation – detection by video camera and acoustic sensors – has proved to be reliable during laboratory testing.
- For on-site testing, the GIL sections should be as long as possible. The length, however, is restricted due to the power of available AC test equipment, or by the requirement of uniform voltage stressing for fast-rise impulse voltages.

It is to note that the knowledge from 40 years of gas insulated technology and the therefore required quality of parts and processes is well understood and reliable diagnostic tools are available. The application of this knowledge to on-site assembly processes for GIL guaranties high quality and reliablity of the GIL in operation.

3.5

Planning Issues

In the following, some information is given about topics which should be evaluated and discussed before a project is executed. The topics are examples taken from experience, but may not cover every topic in each project to be checked. It is recommended to discuss the planning issues of each project with the vendors of the GIL at an early stage. The following topics are chosen as examples.

3.5.1 Network Impact

The network impact of the GIL is part of the network load flow, transient voltages and net frequency studies. These studies are part of the system and network studies, which are necessary to prove the functionality of the planned GIL link in any case and condition during operation or when failures occur. The basis information for the type of GIL investigated in these studies is collected in Table 3.5.1.

The following calculations on net connecting conditions are carried out [110, 118, 124].

3.5.1.1 Net Connecting Rules

The net connecting rules are mandatory requirements defined by the network operator to get connected to the transmission network. Two connecting points are defined at both ends of the GIL. They may be different because of the location and function of the GIL at these points. There could be a one-directional power flow or a two-directional power flow. The rules at these net connection points may vary. One widely used standard requirement is a load factor of 0.95 to 1 in the first quadrant. Higher reactive power is only allowed when the network operator approves.

The net connecting condition of an exemplary network is shown in Figure 3.5.1 [118–120].

Frequencies can vary from 49.5 Hz to 50.5 Hz in a 50 Hz system. Voltage deviations are usually allowed in a tolerance area of about 10%. As shown in Figure 3.5.1, the nominal voltage level of 380 kV has a lower limit of 350 kV and a higher limit of 420 kV for permanent voltages. Above 440 kV the system would shut off for protection.

Gas-Insulated Transmission Lines (GIL), First Edition. Hermann Koch.
© 2012 John Wiley & Sons, Ltd. Published 2012 by John Wiley & Sons, Ltd.

Table 3.5.1   Technical data of the GIL type investigated in the network study [118]

Title	Unit	Value
Nominal voltage	kV	380
Maximum voltage U_m	kV	420
Impulse withstand voltage	kV	1425
Rated current	A	3150
Rated short-time current	kA/s	63 kA/3 s
Rated transmission load	MVA	2200
Overload capability (typical)	%	100
Insulation gas	MPa	N_2/SF_6 mixture at 0.7

3.5.1.2   Load Flow Calculation

The load flow calculation shows the impact of the planned GIL in the network and how the network in a meshed system will change current flow. The basis of these calculations is Kirchhoff's law, where the load flow in a meshed network distributes according to the impedance of the different lines.

The GIL has a very low impedance, and therefore a strong impact on the load flow in the network. These calculations depend on the project conditions and can get very complex. Today, software is available to carry out these calculations and model all the different network conditions [121–123].

To show the principle, a simple example is given below (see Figure 3.5.2). In this example, high-power transmissions from five large offshore wind farms with a total power of 2000 MVA are connected by a 60 km-long GIL from offshore on land to be connected to the network.

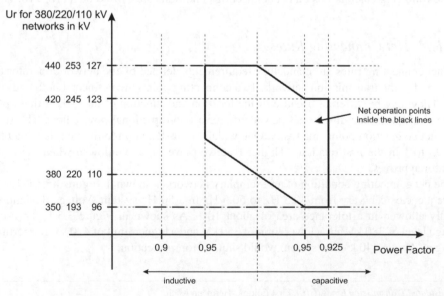

Figure 3.5.1   Reactive power requirements depending on voltage and frequency [118]. Reproduced by permission of © Siemens AG

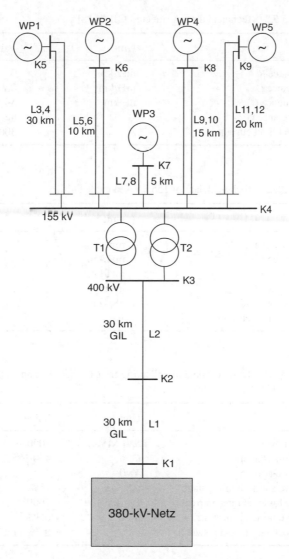

Figure 3.5.2 Connecting a total of 2000 MW from five wind farms over 60 km from offshore collecting point (K3) to the transmission network on land [118]

In Tables 3.5.2–3.5.4 the technical data of the equipment GIL, sea cables and transformers are shown.

The following load conditions have been calculated: full load, half load, no load and reactive power requirements.

3.5.1.2.1 Full Load Calculation Results

The results of the full load calculations (2000 MW) are shown in Table 3.5.5.

Table 3.5.2 Technical data of selected GIL [118]

Title	Unit	Value
Resistance R'	mΩ/km	9
Inductance L'	mH/km	0.215
Capacitance C'	nF/km	54.45
Conductance G'		negligible
Thermal limit current	A	3000

Table 3.5.3 Technical data of 155 kV sea cable [118]

Title	Unit	Value
Cross-section	mm^2	800
Resistance R'	mΩ/km	31.9
Inductance L'	mH/km	0.54
Capacitance C'	nF/km	210
Conductance G'	nS/km	65.97
Thermal limit current	A	775

Table 3.5.4 Technical data of collecting transformer (T1 and T2) offshore [118]

Title	Unit	Value
Rated power	S_{rT} (MVA)	1000
Transmission ratio	$ü$	400/155
Switching group	Yy0	
Relative short-circuit voltage	u_k	15%
Ohmic losses at rated current	P_K (kW)	1000
Relative no-load current	i_1	0.3%
High-voltage side-step switch		±7% in ±7 steps

Table 3.5.5 Transmission losses and reactive power requirement at full load (2000 MW) [118]

Title	Unit	Value
Total losses	MW	23.68
Total efficiency factor	%	98.82
Total reactive power	Mvar	16.48
cos φ net connecting point	ind.	1

Table 3.5.6 Knot identification [118]

Knot	Explanation
K1	Net connecting point on land
K2	At the middle (30 km) of the GIL
K3	At the 400 kV voltage level of the collecting point offshore, GIL side
K4	At the 155 kV voltage level at the collecting point offshore, sea cable side
K5	At the 155 kV collecting point of WP1
K6	At the 155 kV collecting point of WP2
K7	At the 155 kV collecting point of WP3
K8	At the 155 kV collecting point of WP4
K9	At the 155 kV collecting point of WP5

The total losses are 23.68 MW and have a reactive power requirement from the network of 16.48 Mvar for the GIL including the transformers.

The load factor is calculated as $\cos\varphi = 1.0$ at the net connecting point, a very good value.

This fulfils the requirements of the network connection conditions. No additional reactive power compensation is needed. The total efficiency of the GIL including the transformers is 98.82%.

For the calculation of the structure shown in Figure 3.5.2, it has been split into nine knots, as shown in Table 3.5.6.

The calculations at these locations are shown in Table 3.5.7.

The values in Table 3.5.7 show that all voltage levels at each knot are within the accepted values. The voltage increased slightly by 0.54% from the net connecting point on land to the collecting point offshore, reaching 402.17 kV.

In Table 3.5.8 the load flow calculations at full load are shown.

The values in Table 3.5.8 show that sea cables with 746 A and the GIL with 2859 A are not fully loaded.

The currents at both ends of the GIL, I_A and I_B, show almost the same value; this means that reactive power compensation is not necessary. The total losses of the GIL are 13.2 MW, which relates to 220 W/m. When the GIL is laid in a tunnel or directly buried, a thermal calculation needs to be carried out to prove the temperature limits, see Section 3.7.3.3.

Table 3.5.7 Calculated knot voltage [118]

Knot	Knot type	U_K (kV)	$\Delta U_K/U_{nN}$ (%)	δ (°)	P_K (MW)	Q_K (Mvar)	$\cos\varphi$
K1	Slack	400.00	0.00	0.00	1976.32	−16.48	1.0000
K2	PQ	401.17	0.29	1.44	0.00	0.00	−
K3	PQ	402.17	0.54	2.87	0.00	0.00	−
K4	PQ	156.77	1.14	11.31	0.00	0.00	−
K5	PQ	158.65	2.35	13.60	−400.00	0.00	−1.0000
K6	PQ	157.25	1.45	12.09	−400.00	0.00	−1.0000
K7	PQ	156.99	1.29	11.70	−400.00	0.00	−1.0000
K8	PQ	157.55	1.64	12.48	−400.00	0.00	−1.0000
K9	PQ	157.88	1.86	12.86	−400.00	0.00	−1.0000

Table 3.5.8 Calculated results of the load flow at full load [118]

Line	Type	A–B	Length (km)	I_A (A)	I_B (A)	P_v (MW)	Q_v (Mvar)
L1	GIL	K1–K2	30	2853	2855	6.60	–32.86
L2	GIL	K2–K3	30	2855	2859	6.61	–33.20
L3	155 kV	K4–K5	30	746	728	1.60	–41.07
L4	155 kV	K4–K5	30	746	728	1.60	–41.07
L5	155 kV	K4–K6	10	736	734	0.53	–13.52
L6	155 kV	K4–K6	10	736	734	0.53	–13.52
L7	155 kV	K4–K7	5	736	736	0.27	–6.74
L8	155 kV	K4–K7	5	736	736	0.27	–6.74
L9	155 kV	K4–K8	15	737	733	0.80	–20.33
L10	155 kV	K4–K8	15	737	733	0.80	–20.33
L11	155 kV	K4–K9	20	739	731	1.06	–27.19
L12	155 kV	K4–K9	20	739	731	1.06	–27.19

3.5.1.2.2 Half Load Calculation Results

The results of the half load calculation (1000 MW) are shown in Table 3.5.9.

At half load the reactive power increases to 315 MVar, which is an increase of 298 MVar compared to full load. The reactive power from the GIL to the network increases with lower power transmission. This relates to the lower currents and the lower impact of the power transformers. The power load factor decreases to 0.9532 cap. This value is inside the allowed network connecting conditions, as shown in Figure 3.5.1. A reactive power compensation in the half load condition is not needed. The total losses at half load are reduced to 6.38 MW. The efficiency of the total system is 99.36%.

In Table 3.5.10 the analysis of the knot voltages of Table 3.5.6 are shown.

It is shown that all knot voltages are within the required limits. The maximum voltage of the GIL is 403.82 kV at knot 3 at the offshore collection point. This relates to an overvoltage of 0.95%. In Table 3.5.11 the load flow calculations of half load are shown.

The maximum current for the GIL is 1505 A and for the sea cable 402 A. All currents are below their limits. The half load scenario is acceptable with no limitation.

3.5.1.2.3 No Load Results

The results of the no-load calculation are shown in Table 3.5.12.

Table 3.5.9 Transmission losses and reactive power requirements at half load (1000 MW) [118]

Title	Unit	Value
Total losses	MW	6.38
Total efficiency factor	%	99.36
Total reactive power	Mvar	–315.25
cos φ net connecting point	kap.	0.9532

Table 3.5.10 Calculated knot voltages [118]

Knot	Knot type	U_K (kV)	$\Delta U_K/U_{nN}$ (%)	δ (°)	P_K (MW)	Q_K (Mvar)	$\cos\varphi$
K1	Slack	400.00	0.00	0.00	993.62	315.25	0.9532
K2	PQ	402.09	0.52	0.69	0.00	0.00	–
K3	PQ	403.82	0.95	1.38	0.00	0.00	–
K4	PQ	159.03	2.60	5.52	0.00	0.00	–
K5	PQ	160.40	3.49	6.61	−200.00	0.00	−1.0000
K6	PQ	159.32	2.78	5.90	−200.00	0.00	−1.0000
K7	PQ	159.15	2.68	5.71	−200.00	0.00	−1.0000
K8	PQ	159.52	2.92	6.08	−200.00	0.00	−1.0000
K9	PQ	159.77	3.08	6.26	−200.00	0.00	−1.0000

Table 3.5.11 Calculation result at half load [118]

Line	Type	A–B	Length (km)	I_A (A)	I_B (A)	P_v (MW)	Q_v (Mvar)
1	GIL	K1–K2	30	1505	1472	1.79	−69.09
2	GIL	K2–K3	30	1472	1448	1.73	−70.39
3	155 kV	K4–K5	30	402	360	0.45	−48.43
4	155 kV	K4–K5	30	402	360	0.45	−48.43
5	155 kV	K4–K6	10	367	362	0.14	−16.04
6	155 kV	K4–K6	10	367	362	0.14	−16.04
7	155 kV	K4–K7	5	364	363	0.07	−8.01
8	155 kV	K4–K7	5	364	363	0.07	−8.01
9	155 kV	K4–K8	15	373	362	0.22	−24.10
10	155 kV	K4–K8	15	373	362	0.22	−24.10
11	155 kV	K4–K9	20	380	361	0.29	−32.17
12	155 kV	K4–K9	20	380	361	0.29	−32.17

Table 3.5.12 Transmission losses and reactive power requirement at no load [118]

Title	Unit	Value
Transmission losses	MW	0.79
Reactive power compensation losses	MW	0.63
Total losses	MW	1.42
Total efficiency factor		–
Total reactive power	Mvar	−420.03
$\cos\varphi$ net connecting point	kap.	0.0019

Table 3.5.13 Calculated knot voltages [118]

Knot	Knot type	U_K (kV)	$\Delta U_K/U_{nN}$ (%)	δ (°)	P_K (MW)	Q_K (Mvar)	$\cos\varphi$
K1	Slack	400.00	0.00	0.00	−0.79	420.03	−0.0019
K2	PQ	401.92	0.48	−0.04	0.00	0.00	−0.9968
K3	PQ	403.42	0.86	−0.07	0.00	0.00	0.8155
K4	PQ	159.33	2.79	−0.07	0.00	0.00	−0.6026
K5	PQ	160.13	3.31	−0.13	0.00	0.00	0.3775
K6	PQ	159.42	2.85	−0.08	0.00	0.00	0.4479
K7	PQ	159.35	2.81	−0.08	0.00	0.00	0.4549
K8	PQ	159.53	2.92	−0.09	0.00	0.00	0.4363
K9	PQ	159.68	3.02	−0.10	0.00	0.00	0.4204

At no-load condition, when the inductive impact of the GIL and the transformers is the lowest, the highest reactive power compensation is delivered to the network. To meet the net connecting requirements of $\cos\varphi = 0.95$ to 1 inductive, the reactive power of 420.03 Mvar increased losses [124], which results in compensation losses of 0.63 MW and total losses of 1.42 MW.

Besides the highest reactive power delivery to the network at no-load condition, the ferranti effect leads to the highest voltages. In Table 3.5.13 the voltages at the knots are shown and the highest value is at the end of the GIL, at knot 3 with 403.42 kV. This is a voltage increase of 0.86% and is within the limits of the net connecting conditions.

In Table 3.5.14 the load flow calculations at no load are shown.

The load of the GIL and transformers at no load is within the limits.

The operation of the GIL network as shown in Figure 3.5.2 fulfils all net connecting requirements when on land, for the no-load condition a reactive power compensation is installed.

Table 3.5.14 Calculation results at no load [118]

Line	Type	A–B	Length (km)	I_A (A)	I_B (A)	P_v (MW)	Q_v (Mvar)
L1	GIL	K1–K2	30	606	487	0.24	−80.69
L2	GIL	K2–K3	30	487	368	0.15	−82.10
L3	155 kV	K4–K5	30	183	0	0.08	−50.41
L4	155 kV	K4–K5	30	183	0	0.08	−50.41
L5	155 kV	K4–K6	10	61	0	0.02	−16.75
L6	155 kV	K4–K6	10	61	0	0.02	−16.75
L7	155 kV	K4–K7	5	30	0	0.01	−8.37
L8	155 kV	K4–K7	5	30	0	0.01	−8.37
L9	155 kV	K4–K8	15	91	0	0.03	−25.14
L10	155 kV	K4–K8	15	91	0	0.03	−25.14
L11	155 kV	K4–K9	20	122	0	0.04	−33.54
L12	155 kV	K4–K9	20	122	0	0.04	−33.54

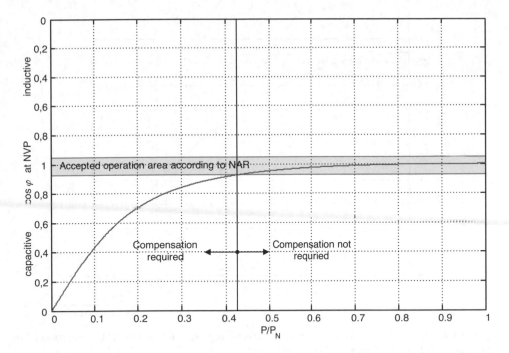

Figure 3.5.3   Load factor at the net connecting point depending on the power: NVP, net connecting point; NAR, net connecting rules in Germany [118]. Reproduced by permission of © Siemens AG

3.5.1.2.4   Reactive Power Compensation

The calculation has shown that the network in Figure 3.5.2 is capacitive in almost any status of operation. Only in the full load condition with 2000 MW is the system slightly inductive.

For the net connecting point, the power relationship is shown in Figure 3.5.3.

The diagram in Figure 3.5.3 shows that below a load factor of 42.3%, reactive power compensation is needed. For compensation, one or several reactor coils or FACTS can be used. The required reactive power can be delivered from one compensation unit at knot K1 on land at the network connection point. The required reactive power values are given in Figure 3.5.4.

The conclusion of this net planning calculation is that GIL can be used for high-power transmission lines (here 2000 MVA per three-phase system) over long distances (here 60 km) without the need for phase angle compensation. Only in the no-load condition is reactive power compensation necessary to fulfil the net connecting conditions. The no-load condition, on the other hand, is not really an operating condition for a high-power transmission line [118].

3.5.2   Reliability

Reliability is a planning issue when the importance of the line and the impact of a possible power supply interruption are evaluated. In principle, the reliability of a GIL can be seen as very high for the following reasons.

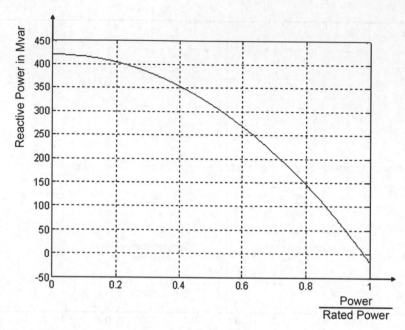

Figure 3.5.4 Reactive power required at the net connecting point K1 on land [118]. Reproduced by permission of © Leibniz Universität Hannover

Experience

Since 1968, gas-insulated technology has been in world-wide use and more than 300 km of pipe lengths are in operation today – in all climatic conditions of Europe, North America, Arabic countries, Asia and Africa. They are installed above ground (about 50%), in tunnels or trenches (about 40%) and directly buried (about 10%); and at voltage ratings of up to 145 kV (about 20%), from 145 kV up to 300 kV (about 20%), from 300 kV up to 550 kV (about 45%) and from 550 kV up to 800 kV (about 5%). Over all these years of operation and installed kilometres there is no major failure reported which led to a power supply interruption. This information has been collected by the IEEE PES Substations Committee on a world-wide basis with the leading manufacturers and users at the table in the working group. The Substations Committee of the Power and Energy Society in IEEE has taken the responsibility on standards, guides, tutorials and panel sessions on GIL [221]. The Substations Committee offers a tutorial on GIL frequently, together with PES conferences.

Principle of Gas Insulation

Another advantage of GIL for high reliability is the principle of gas insulation. The principle of gas insulation is that the insulating gas is free moving and, therefore, allows free movement of electrons and ions. Any kind of electrical load will be transported by convection of the gas to the conductor or ground potential, and will be made ineffective. There is no possibility of storing an electric load at one location and increasing it over the time until the insulation may fail. The solid insulators of the GIL are operated under relatively low electric field strength. The gas insulation defines the dimension of the pipe. The solid insulators can withstand much higher field strength according to the relative dielectric constant of insulators in the range of

$\varepsilon_r = 3\text{--}5$. This means that the electric dimensioning of a GIL is dominated by the gas insulation and the insulators are less stressed.

The principle of gas insulation is one reason for the high reliability of GIL.

No Moving Parts

The GIL has no moving parts and therefore no abrasion. The only elements of a GIL are enclosure, conductor and insulators, no switch or circuit breaker is inside. The only movement comes from thermal expansion when the conductor is heated up by the electric current. These very slow and small movements are captured by a sliding contact system, which guarantees a lifelong operation and no need to open for maintenance.

Non-corrosive

The GIL laid in a tunnel, trench or above ground is protected by its own oxide layer on the surface of the aluminium enclosure pipe. This physical/chemical effect avoids any corrosion of the GIL.

When buried in the ground, active and passive corrosion protection is applied to protect against corrosion.

In conclusion, it can be said that the GIL offers a very high reliability because of its technical design as a gas-insulated transmission line, with no moving parts and corrosion protection [125, 126].

3.5.3 Grounding/Earthing

Grounding or earthing is a substantial safety part of the GIL installation. It covers protection against lightning strikes and touch voltages during normal operation and in case of failure. The GIL system is solidly grounded at both ends and at each structure which holds the GIL, e.g. in a tunnel installation. In a tunnel, above ground or in a trench the metallic enclosure is solidly grounded by connection to a steel structure which holds the GIL. The distances between the grounding points are given by the distances of structures and possible grounding connection in the tunnel, trench or above ground. For each project these distances need to be defined. Typical distances between structures are 20–30 m.

At the end of the GIL a connection is made to the grounding of the substation or to the overhead line tower.

When the GIL is directly buried, the ground connections are made at the underground shafts after each gas compartment length of about 1–1.5 km. Grounding connections are made in the shaft from the GIL enclosure to the grounding of the shaft.

In Figure 3.5.5 it is shown how the GIL is grounded. The metallic enclosure of the GIL is held by a conductive steel structure. The steel structure is fixed to the wall by anchors. The anchors are connected to the steel reinforcement or the concrete tunnel wall. The steel reinforcement of the concrete wall is connected to the grounding/earthing wires in the soil around the tunnel. This principle is used for any accessible GIL enclosure to protect personnel from touch voltages. The GIL is operated as a solid grounded system.

3.5.4 Safety

In principle, the GIL can be seen as safe because of its strong metallic enclosure. In normal operation and in case of an internal arc, no danger is expected.

Figure 3.5.5   Grounding/earthing of GIL. Reproduced by permission of © Alpiq Suisse SA

The GIL has no additional fire load, no external impact if an internal arc occurs and, therefore, no danger for the surroundings. The GIL consists of aluminium, N_2/SF_6 insulating gas and epoxy resin. All these materials are non-flammable under normal conditions. This means that no additional fire load is brought into a tunnel with the GIL.

In case of an internal failure, an internal arc will occur in the GIL. This is the worst case of a failure, which has never happened during the last 35 years and more than 300 km of GIL installed today. A low probability, but even this situation is tested and proved safe. The internal arc will build up pressure, which is kept inside the enclosure; no external impact is seen. The solid metallic aluminium enclosure will keep the hot gases inside the GIL. This means no additional danger is added to the surroundings of a GIL, even inside a tunnel (see Section 3.2).

This high safety feature of the GIL opens up the opportunity of combining a GIL with traffic tunnels (see Section 7.3).

3.5.5 Environmental Limitations

In principle, the GIL fulfils the restrictions coming from the environment. The main environmental limitations are:

- electromagnetic fields
- temperature of the enclosure
- gas-tightness of the insulating gas
- visibility
- agriculture use.

Electromagnetic Fields
Electromagnetic fields play an increasing role in the commissioning process for new transmission lines. Limiting values given by authorities vary strongly world-wide. In Germany, a value of 100 μT, in Switzerland a value of 1 μT and in Italy a lowest value of 0.3 μT are found [158, 161, 162]. Other countries are at some levels in between. Very often, regional requirements are added (e.g., in the USA, individual states and sometimes cities have their own values).

The principle design of GIL, with a solid grounded outer enclosure, offers good opportunities to fulfil the various requirements – especially in densely populated areas where less space and distance is available. The electric field of the solidly grounded GIL is almost zero outside the enclosure pipe and has no external impact.

The magnetic field outside the enclosure pipe of the GIL is small because of the induced reverse current in the enclosure. The negative superposition of both currents (conductor current and induced reverse current in the enclosure) results in a small magnetic field outside the GIL. Only 5–10% of the current in the conductor contributes to the outside magnetic field. The low magnetic field requirements in some countries today are down to 1 μT and can be fulfilled by the GIL, see Chapter 5.

Temperature of the Enclosure
Temperature limitations are given by safety rules for touching temperatures of personnel when tunnel-laid or above ground, and to avoid soil dryout effects when directly buried. Temperature limitations are given in product standards IEC 62271-203 [127] and IEC 62271-204 [54]. The maximum allowed temperature of the enclosure depends on the type of laying, as shown in Table 3.5.15.

When the GIL is laid directly into the ground, the heat transportation is by conductivity of the GIL enclosure pipe and the surrounding soil. To avoid the soil drying out, which is

Table 3.5.15 Maximum temperatures [118]

Type of laying	Temperature
Directly buried	40°C
Tunnel-laid when GIL needs to be touched	60°C
Tunnel-laid when GIL does not need to be touched	70°C

an irreversible process and will increase the thermal conductivity of the soil, special backfill materials are used. These materials provide good thermal conductivity by binding water, and are often mixtures of sand and clay. A better thermal conductivity will increase the power transmission capability to higher currents. In any case, the thermal layout of GIL and the laying conditions need to be calculated for each project. The result of the thermal calculations will define the current rating, the wall thickness of the conductor pipe and the thermal conductivity of the backfill material for the trench. In any case, the long-time maximum temperature must stay below 40°C.

In case of a tunnel-laid GIL the heat transportation from the GIL to the tunnel walls is by convection. The limiting temperature in this case comes from the maximum temperature of the conductor and the insulators. To avoid thermal ageing effects, a maximum enclosure temperature of 80°C is recommended by the international standards.

In some cases, when operational personnel need to touch the GIL, the maximum temperature is limited to 60°C. This value is for safety protection of the operational personnel.

Also, for the tunnel-laid GIL, a thermal calculation is needed for each project. The thermal calculations of a GIL are basic for the project design. When GIL is used within the design limits, no further restrictions need to be taken into account and the GIL can be operated without the risk of thermal ageing.

Gas-Tightness of the Enclosure

To avoid gas releases to the atmosphere, the GIL is completely gas-tight welded and gas handling is done in a closed loop of encapsulated container.

The insulating gases used in the GIL are nitrogen (N_2) and sulphur hexafluoride (SF_6). The majority of the insulating gas is N_2, with typically 80%. N_2 is a natural gas and has no negative impact if released to the atmosphere. Not so SF_6, which is a global warming gas and should not be released to the atmosphere. Also, SF_6 is relatively expensive compared to N_2.

The gas-tightness is also needed for electric integrity of the system.

For both reasons, the GIL concept is a completely welded joint technology with gas-tight welds. These welds are produced by automated welding machines and controlled by automated ultrasonic weld control. For long distances the GIL is separated into gas compartment sections of about 1 km length. This limits the risk of high gas losses in case of damage of a GIL enclosure pipe.

Visibility

High-voltage overhead transmission lines have great public opposition. Commissioning processes can last for years or even decades, to get the permission to build. When the power transmission line is underground like the GIL, then public acceptance is higher.

A tendency in the future with large-scale regenerative energy generation will be to ask for short construction times to connect a wind farm or solar thermal power generation plant. The higher cost of the GIL may be justified against the long commissioning time and loss of income by not selling electricity on the market.

Visibility has a stronger role in densely populated areas or in areas with environmental protection.

In some cases it might be of advantage to combine overhead lines with GIL, which can be done simply by bushings including overvoltage surge arresters. With such combinations, overhead lines may be built in rural areas where no strong public opposition is expected and the underground GIL can be used in more populated areas or environmentally protected areas.

Agriculture Use

After the GIL is laid in the ground, the soil above the ground can be used for farming. No large trees should be allowed to grow on top of the GIL, to allow accessibility in case of repair.

This limitation on agriculture use has been applied with oil and gas pipelines for several years now, with good experiences. Land owners are compensated during the construction time, to allow all the construction site traffic. Later compensations are given when access is required for operation, maintenance or repair.

The main impact on the landscape is during the construction phase. To open a 3–5 m-wide trench for one or two GIL systems requires moving large amounts of soil, storing it and filling it back. The construction time is limited; a 1 km section will need 3–6 months depending on the accessibility of the site and the site conditions of materials (sand, clay, rock).

A long-distance GIL will be split into sections of about 1–2 km and, therefore, the construction site opening is of the same length. In most cases there will be no construction site with open trenches of tens of kilometres.

3.5.6   Electric Phase Angle Compensation

The capacitance of the GIL is about two to three times higher than an overhead line, and about four to five times smaller than a solid insulated XLPE cable. A 400 kV GIL has phase angle compensation requirement when reaching lengths of 60–80 km. The exact length depends on the network condition at both ends of the GIL, and needs to be calculated with a network study.

As found in Section 3.5.1, the compensation requirement is at no-load condition and relatively low MVAr are needed. Depending on the network conditions at the connecting points, the compensation requirement needs to be calculated.

3.5.7   Loadability and Capability/Overload

3.5.7.1   General

Loadability and overload of GIL are important operational features to adapt to network transmission changes. The GIL, by its principle design of gas insulation inside and large diameter of the enclosure pipe, has a good thermal behaviour. The heat produced by the conductor current is transported by gas convection to the enclosure pipe and then by the relatively large surface area to the surroundings.

In the case of tunnel, trench-laid or above ground, the heat goes by convection to the surroundings. When laid in soil, the heat is transported by conductivity to the soil.

In both cases the heat transport is fast and thermal heat-up of the GIL is slow. In case of the tunnel-laid GIL the time constant is in the range of some hours, depending on the surrounding air conditions. In case of buried GIL the time constant can be as long as some days before the thermal limitations are reached, depending on the conductivity of the soil surroundings of the GIL. Besides the surrounding conditions of ambient air or soil conductivity, the electrical transmission conditions define the overload capabilities. See also Section 3.7.2 for thermal stress when tunnel-laid and Section 3.7.3 for thermal design when directly buried.

This can cover short time overload conditions of the GIL without risk of over-temperatures. The calculations can be made when the thermal constant of the GIL in its laying condition is known.

3.5.7.2   Calculating Overload for Ambient Temperature

These calculations are only applicable for tunnel or trench-laid or above-ground GIL. In case of directly buried GIL, refer to Section 3.7.3.

The need for temporary overload may come from unanticipated load or generation growth or from system reconfigurations. The GIL temperature limits are given by design and have been verified under type conditions for rated operation conditions. External impact from ambient air, soil conductivity and solar radiation needs to be considered.

The thermal overload formulas are given in IEC 62271-1 [82], the dimensioning of the GIL is given in IEC 62271-204 [54].

The maximum allowed temperatures are related to the different components of the GIL.

Insulating Gas
The insulating gases used in GIL are SF_6 and N_2. These gases are stable and can be used up to 500°C. So, there is no practical limitation on the GIL from the insulating gases.

Insulators
The insulators are made of cast epoxy. There are different types of cast epoxies used, depending on the manufacturers. The maximum temperatures of these cast epoxy insulators are between 110°C and 125°C. These values are related to the insulator materials and are standardized in IEC 62271-204 [54] or IEEE C37.122 [128].

Sliding Contacts
Sliding contacts are usually silver plated and are used to compensate the thermal expansion of the conductor. The temperature limits of such contacts are limited to 105°C. This value is a standardized value from IEC 62271-204 [54] or IEEE C37.122 [128].

The main influencing parameter for the thermal layout is the diameter of the enclosure pipe. With the diameter, the surface is fixed and the possibility of transferring heat to the surroundings.

The thermal time constant is related to the diameter of the GIL and is measured during the heat-run test of the type tests. For the GIL under standard ambient air of 40°C, this thermal time constant is typically to 3 hours. To calculate the heat increase related to a load increase, the pre-load and post-load conditions must be specified. Typically, load cycles of 2 or 3 hours before and after the peak load are evaluated. To limit the peak solar load, sun shields may be used when the GIL is installed above ground.

Calculation of dynamic load conditions is complicated. In most cases the overload situation is limited to an overload condition of 1 to 4 hours. The thermal overload calculation as given in IEC 62271-1 [82] covers continuous or temporary overload at changing operating conditions. In Table 3 of IEC 62271-1 the maximum temperature rise limits are given, which shall not be exceeded at any part of the GIL. With these limiting values the GIL can be assigned higher than rated normal currents for temporary overload values without reaching the limiting temperature values. This is the case, for example, if the real ambient temperature at the operating location of the GIL is lower than 40°C.

The continuous and temporary overload calculations are based on values obtained from temperature rise test and test parameters: rated currents, thermal time constant, temperature rise, ambient air temperature and maximum operating temperatures. The following equations apply for calculating overload capacities.

The allowable continuous current (I_s) for a given ambient temperature θ_a:

$$I_s = I_r \left[\frac{\theta_{max} - \theta_a}{\Delta\theta_r} \right]^{\frac{1}{n}}$$

The operating temperature during overload:

$$\theta_s = \Delta\theta_r \times \left(\frac{I_s}{I_r} \right)^n \times e^{-t} \tau + \theta_a$$

or, the allowable duration (f_s) of temporary current I_s after carrying a current I_i:

$$t_s = -\tau \ln \left[1 - \frac{(\theta_{max} - Y - \theta_a)}{\left(Y \left[\frac{I_s}{I_i} \right]^n - 1 \right)} \right]$$

$$Y = (\theta_{max} - 40) \times \left[\frac{I_i}{I_r} \right]^n$$

where
θ_{max} = maximum allowable total temperature (°C) according to Table 3
θ_a = actual ambient temperature (°C)
$\Delta\theta_r$ = temperature rise at normal current I_r
I_r = rated normal current (A)
τ = thermal time constant (h)
n = overload exponent taking into account material, heat radiation, convection, etc.
I_i = initial current before application of overload current (A)
I_s = overload current (A)
t_s = permissible time (h) that overload current (I_s) can be carried without exceeding maximum temperatures allowable (θ_{max})

In general, no additional temperature rise tests are required if an exponent $n = 2$ (as a conservative estimate) is used for the determination of the operating temperature during overload or allowable overload duration. An exponent lower than $n = 2$ may be used for the calculation of the overload rating. It has to be demonstrated by calculation from test data.

In Figure 3.5.6 the transient overload is shown for the case that at 40°C ambient temperature a lower pre-overload current (about 50% of the rated continuous current) is changing into an overload current (about 10% higher than the rated continuous current) and finally falling back to the rated continuous current.

The temperature shows an exponential heating curve after the overload current applies until it ends.

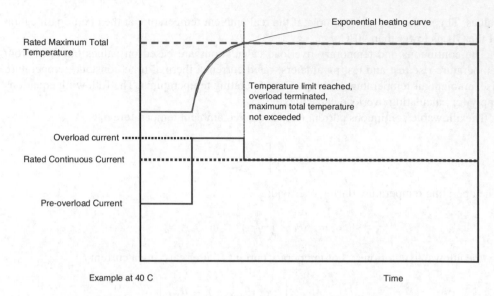

Figure 3.5.6 Transient overload. Reproduced by permission of © ILF

In the case of lower ambient temperature around the GIL, the allowed possible temperature rise for overload is larger.

As shown in Figure 3.5.7, the rated maximum total temperature of the GIL allows additional heat above the ambient temperature of 40°C. If the ambient temperature is only 0°C, the additional heat coming from the overload is the difference in I^2R of the total temperature minus the 40°C temperature rise.

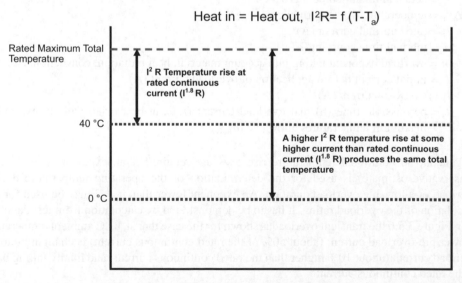

Figure 3.5.7 Rated continuous current adjustment for ambient temperature. Reproduced by permission of © ILF

3.6

Specification Checklist

GIL projects are specific, and a specification checklist can only give some framing information on possible investigations. The local conditions for each project are different, and a big variation of rules coming from authorities needs to be fulfilled. A detailed route planning is required to find all cost impacts from obstacles like river crossings, railroad crossings, highway crossings, the type of soil, accessibility and landscape conditions [221].

Route Planning

The route planning is an optimization process with several interactive steps and possible alternatives. First, the soil conditions and the width for the right of way for directly buried GIL or for tunnel or trench-laid GIL. Public housing, living areas, industrial areas, all kinds of infrastructure including streets, pipelines, water canals and any other item in the way of building needs to be identified. The ownership is of great importance, public or private, and will have implications for the commissioning process.

Availability of planning data for soil and underground installations: good data will help to avoid big surprises, while bad planning data may cause high additional costs when obstacles are detected during excavation. Accessibility of the route during the construction time is an important planning input. Maximum transportation lengths and weights define the assembly process, and have a direct impact on the total project costs. As a simple reminder: the longer the transportation lengths of the GIL section, the lower the number of on-site weldings and the lower the project costs!

Site Conditions

The GIL is assembled on-site. This means that the single elements to assemble the GIL are transported separately from their pre-manufacturing location to the site assembly location. This requires site conditions which allow the set-up of an assembly tent to store pre-delivered elements and access roads which can handle truck traffic. These pre-assembly sites need to be close to the GIL route and will be required at distances of a few kilometres along the route (see Section 3.3.12).

Gas-Insulated Transmission Lines (GIL), First Edition. Hermann Koch.
© 2012 John Wiley & Sons, Ltd. Published 2012 by John Wiley & Sons, Ltd.

Technical Data

The rated voltage and current are the main technical data necessary to calculate the temperature along the route and fix the test voltages in accordance with the standards [54]. The size of gas compartments will be specified, together with the route planning and the possibility of local underground shafts for length limitations of the gas compartments (which should be in the range of 1–1.5 km). See Section 3.7.

Underground Shafts

Along the route at distances of 1–1.5 km, underground shafts are needed for the disconnecting unit when the GIL is directly buried. These disconnecting units are underground and accessible for inspection when necessary. They may hold monitoring devices. See Section 3.7.

Monitoring

For monitoring of the GIL a parallel data line to connect each underground shaft to the operation and control centre is needed. Typically, this is done by optical data cables. See also clause 3.3.13.

Engineering Studies

Each route needs to be engineered and the following studies may be needed:

- Network planning and simulations.
- Transient voltages on the line coming from lightning overvoltages or switching overvoltages.
- Surge arrester concept to protect the GIL.
- Electromagnetic field calculations to prove limit values.
- Seismic calculations, when required.
- Soil and ground studies.
- Grounding (IEC term)/earthing (IEEE term) of the GIL and the secondary systems.

Site Logistic Studies

- Transportation length and road or rail accessibility.
- Storage of material – mainly, enclosure and conductor pipes are large (12–18 m typical) and need much space.
- On-site transportation of pre-assembled GIL sections.
- Quality insurance system on-site.
- Testing procedures on-site.

Gas Zone Specification and Identification

The GIL is separated into gas compartments or gas zones. It is important to the operator of the GIL to know exactly at what location the gas compartments are separated and gas zones are formed.

The typical gas zones of a GIL are:

- Straight GIL unit of up to 1 km length.
- Disconnecting unit.
- Bushings to overhead lines.

The gas-tight insulators should be indicated from the outside by colour or by symbol.

The gas zone schematic should give an overview of the total transmission line including the following information:

* Gas compartments.
* Valves for filling and vacuum pumping.
* Density monitors for control signals.
* First alarm levels of density (usually 5–10% below the filling density).
* Second alarm level before minimum gas density will trip the GIL.
* Gas monitoring system for indication in the control centre.

Environmental Restrictions
* Neighbours of line.
* Aesthetics for public.
* Soil movement.
* Sound and noises.
* Electromagnetic fields (EMF).

Standards and Regulations
* IEC Standards
  o IEC 62271-204 'GIL' [54]
  o IEC 62271-1 'Common Clauses' [82]
* IEEE Standards
  o IEEE 1677 'GIL Guide' [129]
* CIGRE Brochures
  o CIGRE TB 260 'N_2/SF_6 Gas Mixtures' [130]
  o CIGRE TB 218 'Gas Insulated Lines (GIL) [131]
  o CIGRE TB 351 'Application of Long High Capacity GIL in Structures' [132]
* Regulations are released by authorities and vary from country to country.

Tests and Inspections
* General inspection of the assembly.
* Control system function test.
* HV and partial discharge tests on-site.
* Gas-tightness check integrated in the orbital welding process.
* Control system function test.
* Arc location system function test.
* Nameplate check.

Project Deliverables
* Route plan with all underground shafts.
* Grounding (IEC term)/earthing (IEEE term) plan.
* Control schematics for gas monitoring and arc location system.
* Instruction manuals.
* Gas zone diagram.

- Physical drawing.
- Section view of GIL.
- Static and dynamic load plan.

Local Specific Requirements
- Reliability criteria.
- Compliance with local authorities.

3.7

Laying Options

3.7.1 General

In this section, the different options to install or lay a GIL are explained.

Most of the laying options have been developed parallel to requirements coming from the different applications. First, GIL installations have been placed inside substations to connect different parts of the high-voltage side of the substation. Here, laying options of above-ground installations using steel structures of different heights or laying in concrete trenches in the ground were used. In some cases GIL was laid similarly to pipelines, directly in the soil, with a corrosion protection on the outside and covered with soil.

In applications like hydropower plants with large dams or tunnels inside a mountain where the generators are located, GIL was installed in tunnels.

In each case, specific requirements on mechanical and thermal stresses need to be covered and the advantages of one or other laying option need to be evaluated.

GIL is a solid metallic pipe and with this, its physical specifics are fixed.

Mechanical Solid
The GIL is a solid pipe which needs only a few support structures over long distances, e.g. 20–40 m. The distance of the support structure is related to the pipe diameter and wall thickness, which makes the mechanical stiffness.

Elastic Bending
The aluminium pipes of the GIL can be bent down to a bending radius of 400 m. With this bending radius, the curve of most landscape profiles can be followed.

Angle Elements
The GIL can easily be combined with angle elements. This allows any angle, even 90° directional changes in a spot, and is often used in above-ground installations.

These possibilities allow a great variation of applications. Table 3.7.1 gives an overview.

The GIL offers many laying options for high-voltage, high-capacity transmission lines. With future requirements coming from the connection of remote, regenerative energy resources, the GIL will also offer new possibilities for densely populated or environmentally protected areas.

In the following, the different laying options will be explained in more detail.

Gas-Insulated Transmission Lines (GIL), First Edition. Hermann Koch.
© 2012 John Wiley & Sons, Ltd. Published 2012 by John Wiley & Sons, Ltd.

Table 3.7.1 Laying options of GIL

Laying option	Field of application
Above ground	Inside substations or power plants.
	Using steel structures of 6 m height or even higher.
Trench-laid	Inside substations or power plants.
	Using concrete trenches to cross other lines.
Tunnel-laid	At hydropower stations or dams.
	Under densely populated cities: Electric Energy Tunnel.
Directly buried	Inside substations or across country.
	An underground transmission where overhead lines cannot be built.
Combined with bridges	Attached to or inserted into bridges.
	The high safety of GIL allows combinations with bridges.
	They need to be considered in the planning stage.
Combination with traffic tunnels	Addition to a road or railroad tunnel is possible due to the high safety of GIL.
	To be considered in the early planning stage.
Combination with highways	Added beside or under the highway.
	The GIL can be used because of low electromagnetic fields and high safety.

3.7.2 Above-Ground Installation

3.7.2.1 General

Today, most of the world-wide installed 300 km of GIL are installed above ground on steel structures in power plants or substations (more than 50%). In substations it is often required to cross other high-voltage lines to connect, for example, the incoming or outgoing overhead line with switchgear of the substations. When overhead lines cross, the height increases significantly and maybe permission will not be obtained.

Alternatively, in gas-insulated substations using GIS it is necessary to expand the distance between the phases of the GIS to about 1 to 4–5 m for the overhead line, e.g. 400 kV. Here, GIL is often used.

Or, in some countries with high air pollution, connections are made from the machine transformers of the power plant to the switchgear building by GIL (e.g., Saudi Arabia or other desert countries).

The use of aluminium pipes for the enclosure under atmospheric conditions develops an oxide layer, which protects the aluminium from corrosion. This simplifies the use of GIL above ground. These above-ground installations may be low and close to the ground or even very high to give room for underpassing the line with truck traffic in the substation.

In some cases, both laying options are used in one power plant or substation depending on the specific requirements.

Laying Close to Ground
The laying option close to the ground needs the smallest amount of steel for structures to hold the GIL. In Figure 3.7.1 such a steel structure is shown at the right side of the photo, including housing for the gas density monitoring fixed to the same steel structure [133, 134].

Figure 3.7.1   Above-ground laying close to the ground. Reproduced by permission of © Alstom Grid

Laying High Above Ground

Typical laying heights for this laying option are 5–8 m above ground. This allows traffic below the GIL without restrictions by high-voltage requirements, but needs the most steel for structures to hold the GIL. In Figure 3.7.2 the connection of the GIL to the switchgear building is shown. In Figure 3.7.3 the connection of the GIL to the overhead line is shown.

Crossing of Systems

To cross out three-phase systems when exiting a switchgear building with GIS inside is an often-used application for GIL. In Figure 3.7.4 such an application is shown.

To connect transformers by bushings, similar steel structures are used.

3.7.2.2   Corrosion Protection

When aluminium is exposed to air, the oxygen will create an oxide layer, only a few nanometres thick, but very hard and very dense so that no further corrosion is possible. Aluminium is self-protecting. Only when water penetrates for a very long time (several weeks or months) may the oxide layer get dissolved by chemical reactions with chlorides, and then may punctually

Figure 3.7.2 GIL connection to the switchgear building with high above-ground steel structures. Reproduced by permission of © EGAT and © Siemens AG

Figure 3.7.3 GIL connection to the overhead line with high above-ground steel structures. Reproduced by permission of © EGAT and © Siemens AG

Figure 3.7.4   Welded connections on above-ground steel structures. Reproduced by permission of ©
CGIT Systems

corrode. To avoid such an impact, which can cause corrosion, protection is needed by coating
when directly buried or maybe a roof to disroute water in a tunnel.

Rain and fog will not harm the oxide layers, except directly at the sea (salty air) or close
to a desalinating power plant (salt in the air). In most applications world-wide no additional
corrosion protection is needed, in some few cases a paint or coating is used. The surface of an
aluminium enclosure pipe with oxide layer is shown in Figure 3.7.5.

Figure 3.7.5   Oxide layer at the enclosure pipe. Reproduced by permission of © Alpiq Suisse SA

Figure 3.7.6 Connection of GIL sections on high above-ground steel structures. Reproduced by permission of © Alstom Grid

3.7.2.3 Mechanical Stress

The solid structure of the GIL requires only a few steel structures to support the GIL, as shown in Figure 3.7.6. The GIL sections are connected by joints on-site when laid on steel structures. The steel structures are fixed to concrete sockets. The GIL sections are prefabricated, with section lengths of 10–20 m.

3.7.2.4 Thermal Stress

Above-ground installations are exposed to ambient air and sun. In cold countries the possibility of ice load and wind load needs to be considered. Ice may add a large amount of weight to the steel structures and wind may add bending forces.

In hot and sunny countries the solar radiation will add heat to the GIL, which needs to be considered when the current rating is defined. See also Section 3.5.7 on loadability and overload. In tropical and desert countries it may be necessary to shade the GIL by small roofs.

In any case, the thermal expansion of the GIL pipes needs to be considered. From the cold condition of the GIL when power transmission is switched off to the hot-day, full-load condition, the thermal expansion joints need to take care of the thermal expansion. For a 100 m-long section and a temperature increase or decrease of 40°C, the thermal expansion may result in a length extension of about 100 mm.

Longitudinal or angle thermal expansion joints are part of the above-ground GIL installation.

3.7.2.5 Evaluation of Above-Ground GIL

The above-ground installation of GIL is mostly used in substations and power plants in non-accessible areas to the public. A fence will protect the substation or power plant. The above-ground installation is easy, and offers maximum flexibility for high-voltage connections. Especially in the case of crossing overhead lines, GIL is used in substations or power plants. With the GIL an existing overhead line or bus bar can be crossed, because the GIL has ground potential at the outside. The above-ground installation is easily accessible and view inspections can be done at any time.

3.7.3   Trench-Laid

3.7.3.1   General

The reasons for trench-laid GIL are similar to those for above-ground installations. It is the crossing of overhead installations in substations and power pants when the route underground is a better solution.

Trenches are usually made of concrete, and the GIL is fixed on small steel structures inside. The concrete trench is more expensive than above-ground installations, because of the digging underground, but less expensive than accessible tunnels. The trenches are covered with concrete blocks. Trenches are made for one three-phase system with three GIL pipes (as shown in Figure 3.7.7), or can hold two three-phase systems with six GIL pipes.

Figure 3.7.7   GIL in a trench. Reproduced by permission of © CGIT Systems

3.7.3.2   Corrosion Protection

The corrosion protection for trench-laid GIL is the same as for above-ground installations. The trench needs to be free of permanent water to avoid the distortion of the oxide layer of the aluminium surface. Temporary water is permissible for a short time (some days). Then the surrounding needs to be dry again. This is usually done by a dewatering and drainage system.

Care needs to be taken in aggressive ambient atmospheric conditions, e.g. in chemical plants or in seawater desalinating plants. In any case, the prior solution is to keep the trench dry. If painting is needed, control and maintenance are required. If the conditions in the trench cannot be kept dry, corrosion protection similar to the directly buried GIL will be required.

3.7.3.3   Mechanical Stress

The mechanical stress on the GIL in a trench is similar to above-ground installations. Steel structure distances are the same. There is no additional force coming from ice load or wind load.

3.7.3.4   Thermal Stress

The thermal stress of trench-laid GIL is similar to the above-ground installation of GIL, except there is no solar radiation.

Different from above-ground installations there is no wind for cooling, only thermal convection of the ambient air in the trench. Thermal calculations at rated currents, the ambient air temperature and thermal convection in the trench are required in any case.

3.7.3.5   Evaluation

Laying in a trench is usually the solution if the above-ground solution is not possible or too complex. The higher cost for the trench compared with an above-ground solution needs to be justified. The trench solution allows (as the above-ground solution when installed) high, free truck traffic in the substation or power plant.

In some cases, trenches are also used outside substation fences, then access to the public needs to be avoided by covering the trench.

3.7.4   Tunnel-Laid

3.7.4.1   General

The tunnel-laid option is a very compact laying method with typically two GIL systems in a small space of 2.5 m². Such a tunnel is accessible to personnel and the GIL remains accessible after laying. This allows easy control and in case of repair, easy access. Tunnels may be built in an open trench in segments, later covering with soil (minimum coverage height of 1 m). Or the tunnel may be bored to deeper depths, with accessibility by shafts. Bored tunnels are round and have a diameter of 3–4 m for two GIL systems.

So-called microtunnels of diameter about 3 m are today economical solutions in cases when directly buried GIL are not possible – for example, in urban areas, or in crossing mountains, or in connecting islands under the sea. These microtunnels are usually the shortest connection between two points and, therefore, reduce the cost of transmission systems. After commissioning, the system will easily be accessible [6, 135].

Tunnels are ventilated and can have forced ventilation to increase the cooling and with this the power transmission capability. Forced cooling with refrigerated air is usually not required, and should be avoided because of high maintenance costs of the cooling system. Corrosion protection and mechanical stresses are similar to the above-ground laying options.

3.7.4.2   Open Trench-Laid Tunnel

For the open trench tunnel laying option a trench will be opened from the top and the tunnel segments will be laid in the open trench and finally covered with one to two metres of soil. The tunnel segments can be produced on-site using concrete, or the segments are prefabricated and assembled on-site (see Figure 3.7.8).

With this laying option the landscape will be recovered after the construction works are finished. Full agriculture use is possible. It should be protected from large plants like trees to keep the tunnel accessible.

In Figure 3.7.9 a view into the tunnel laid in an open trench at Geneva Airport is shown. The two GIL systems are fixed at each side of the tunnel, on steel structures held by the tunnel walls. In the centre, space for a walkway is allowed (about 0.8 m width).

Such tunnels can be built in densely populated areas under or beside streets and highways.

The required space for two GIL systems is about 2.5 m^2. The tunnel is seen as a non-accessible tunnel in terms of required service inspections, which are not necessary for normal

Figure 3.7.8   Open trench tunnel laying option (1) Delivery of GIL units, (2) assembly and welding, (3) pulling into the tunnel, (4) high-voltage test. Reproduced by permission of © Siemens AG

Figure 3.7.9 View into the tunnel at PALEXPO with one, three phase system of GIL installed. Reproduced by permission of © Alpiq Suisse SA

operation. Only over longer time sequences, e.g. each month, will it be necessary to view check the tunnel. Also, no additional cooling will be required for rated power transmission – natural convection of air in the tunnel will be sufficient to transport the heat to the surrounding tunnel walls and into the soil.

For assembly, the GIL as shown in Figure 3.7.10 will be jointed by welding at the tunnel entrance and pulled into the tunnel. By elastic bending, the jointed GIL will follow the tunnel curvage while sliding or rolling over the support structure. The bending radius seen in the photo is about 700 m.

3.7.4.3   Bored Tunnel

The developments in tunnel boring have brought about more speed and higher accuracy. This has lowered the cost, so that more bored tunnels are built today. GIL is added to such tunnels by fixing to the tunnel walls. Two GIL systems can be installed, one at each side (as shown in Figure 3.7.11). In the centre of the tunnel a walkway is available for easy access.

The diameter of this round tunnel is 3–4 m, depending on space requirements for the accessible walkway. The steel structures carry the weight of the GIL and give guidance when curves are made by the tunnel down to a bending radius of 400 m.

To install the GIL in a tunnel, GIL segments of about 12–14 m length are pulled into the tunnel through shafts after they have been connected by welding [136]. The support structures can have distances of 20–40 m.

Bored tunnels are usually deep underground. The minimum depth is three times the tunnel diameter, which is 12 m for a 3 m tunnel diameter. But they can be as deep as 20 or 30 m under cities. Access to the tunnel is by shafts, as shown in Figure 3.7.12. The GIL segments are brought into the tunnel by cranes through the shaft and may have lengths of 12–14 m, depending on the shaft diameter. The shaft diameter is given by the type of boring machine

Figure 3.7.10 GIL in a tunnel with 700 m bending radius. Reproduced by permission of © Alpiq Suisse SA

when the tunnel is made. The prefabricated tunnel segments are brought to the shaft by trucks (1). A crane will bring the GIL segments down to the joint point, where a welding tent is placed (2). The distances between shafts are several kilometres and the GIL is pulled into the tunnel after welding (3). After the GIL has reached the next shaft, a high-voltage test of this section will be made through the shaft (4).

In the tunnel, the GIL is laid on slides or rollers to allow thermal expansion during operation. The rollers are held by fixing structures, see Figure 3.7.13.

Figure 3.7.11 Bored tunnel inner wall made of concrete segments. Reproduced by permission of © Alpiq Suisse SA

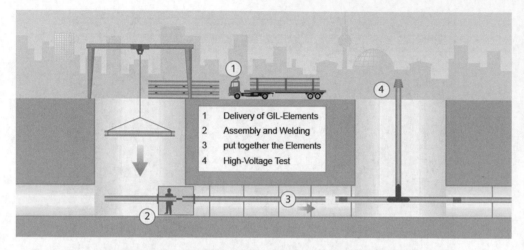

Figure 3.7.12   Tunnel laying through a shaft: (1) Delivery of GIL units, (2) assembly and welding, (3) pulling into the tunnel, (4) high-voltage test. Reproduced by permission of © Siemens AG

Figure 3.7.13   Rollers to hold the GIL on the structure. Reproduced by permission of © Siemens AG

Figure 3.7.14   View of the assembly area. Reproduced by permission of © Siemens AG

In Figure 3.7.14 the truck transportation of a set of six GIL segments to the shaft is shown. The shaft is covered by an assembly tent to protect the assembly and welding area from rain and other severe ambient conditions [100, 152].

3.7.4.4   Corrosion Protection

When the GIL is laid in a tunnel, the ambient air will create an oxide layer. This oxide layer is very thin (only nanometres) and will stop corrosion. No further coating or passive corrosion protection is needed. Only when permanent water drops from the wall penetrate the enclosure pipe – including with chlorides – might the oxide layer be destroyed and punctual corrosion occur. Then the GIL in a tunnel needs protection against permanent water dropping, e.g. by a small roof.

3.7.4.5   Mechanical Stress

The mechanical stress of a GIL in tunnels is similar to above-ground installations. Steel structures are attached to the tunnel walls to hold the GIL pipes. The tunnel walls carry the load from the surrounding soil and the fixing points need to be coordinated to the mechanical stability of the tunnel. As the lightweight aluminium GIL does not bring large forces to the tunnel walls, this is usually not a problem.

The ambient conditions are different from above ground. There is no sun radiation and humidity is normally high in tunnels; care needs to be taken to avoid water condensation and the drainage system needs to take care of water entering the tunnel. The steel structure will also take care of the thermal expansion of the GIL when heated up or cooled down during operation.

3.7.4.6   Thermal Stress

Thermal stress in the tunnel is similar to above-ground installations. The heat produced by the transmitted current needs to be transported to the tunnel and, with natural or forced ventilation, to the outside through the shafts and to the tunnel walls with thermal conductivity to the surrounding soil. For each application a thermal study is required to meet the limit temperatures of the GIL, the tunnel and the surrounding soil at a given power transmission capability.

3.7.4.7   Evaluation

Tunnel installations are high investments on the one hand but they also open up the opportunity of high and compact electric power transmission without being visible from above. The cost of tunnels depends very much on the way they are built. From simple close-to-the-surface tunnels, to water-tight, deep-under-ground tunnels the cost variation is high. A case-by-case investigation of the best solution is necessary at the project planning phase. Tunnel-building technology is constantly improving, and tunnel costs are going down. For the GIL, the tunnel-laid version is very convenient as no corrosion protection is needed and the GIL stays accessible.

3.7.5   Directly Buried

3.7.5.1   General

Besides the tunnel-laid GIL, the directly buried GIL is becoming more important in use for undergrounding power transmission. The directly buried GIL is strongly related to pipeline laying, which is used for oil or gas pipelines. This laying method is a continuous process in an open trench and the goal is to cover long distances as fast as possible. The directly buried GIL follows this method.

The directly buried GIL is the electric power transmission option for long-distance applications. At high energy transmission levels of, for example, 3000 MW per system and over distances of 100 km or more, the directly buried GIL will be an "electrical pipeline". The GIL laying methods are adapted to oil and gas pipeline laying techniques where the goal is to install long lines in a short period of time. Gas pipelines are world-wide a proven technology for long-distance transportation of gas for energy supply. In Europe and North America, thousands of kilometres are in service without technical problems or non-acceptance by the public. The basic feature is a continuous laying process, at parallel construction sites along the line [137–139].

Pipelines usually use steel as an enclosure pipe. The pipes are jointed by arc welding in a continuous process and then laid into the trench. The GIL uses AC transmission voltages in aluminium pipes. The joints are welded by arc welding. The mechanical forces of the pipes need to cover mechanical stress during laying and gas pressure load, soil load and thermal forces during operation. In this section the correct dimensioning of the directly buried GIL is explained.

Directly buried GIL is an economical solution and offers fast laying for cross-country transmission lines. Similar to pipeline laying, the GIL is continuously laid into an open trench with a nearby pre-assembly site to assemble the GIL sections. This reduces the transportation of GIL units on-site. With an elastic bending radius of 400 m, the GIL can follow the contours of the landscape.

Laid in the soil, the GIL is continuously anchored, so no additional thermal expansion needs to be compensated. Four standard units of GIL make the system economical to install on-site, and reach short erection times to cover long distances [102, 140].

The enclosure pipe is corrosion protected by a polymer coating. The coating technology is taken from the gas and oil pipelines, where good experiences have been achieved. Also, the angle elements have been corrosion protected with a coating. The joint area, where on-site orbital welding joints the GIL sections, has a special on-site coating to protect the weld.

The directly buried laying option is typical for open landscape across country. It may be in connection with existing pipeline routings, or with highways, or with railroads, or in the open field. After installation, the minimum coverage of soil is 1 m so that the fields above can be used for agriculture.

Directly buried GIL are not visible and access is only needed at distances of 1–1.5 km to underground shafts. Therefore, only a manhole is needed. Once installed, no operational or maintenance access is necessary.

Overhead lines see increasing public opposition, and to erect new overhead lines close to residences or in protected landscapes or national parks is almost impossible. The commissioning process can take decades before a compromise can be found with the public. The required restructuring of the electric power transmission network towards an all-renewable electric power generation will require new high-power transmission lines on a short time scale. GIL directly buried is one possible solution.

3.7.5.2 Laying Process

Similar to pipeline laying, the GIL will be continuously laid with an open trench using a nearby pre-assembly site to reduce the transportation of GIL units. With the elastic bending of the metallic enclosure, the GIL will flexibly adapt to the contours of the landscape. In detail, different landscapes need different technical solutions.

In the soil, the GIL is continuously anchored, so that no additional compensation elements are needed. In Figure 3.7.14 the laying procedure for directly buried GIL is shown [34, 36, 79, 102].

The construction process is shown in Figure 3.7.15. Single GIL units of 12–14 m length are prefabricated in mobile prefabrication tents (1) close to the location of laying. Inside the welding tent (2), which is direct at the trench or in the trench, the GIL units will be jointed by welding. Sections of 100 m length or even longer may be jointed beside the trench before laying into the trench. Another laying method is to use the disconnecting shaft (4) to locate the welding tent on top and to pull the GIL sections into the open trench. This laying method combines the tunnel laying pull-in method with the open trench laying directly into the soil.

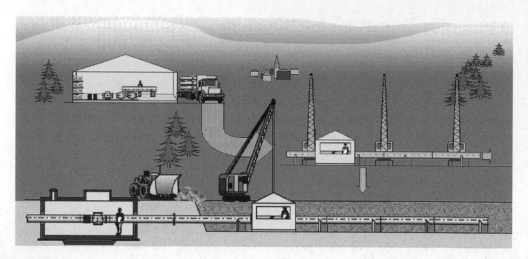

Figure 3.7.15 Construction process of a directly buried GIL. Reproduced by permission of ©
Siemens AG

The welding in the trench will be done inside a welding tent in the trench (3). Sections of
1000 m to 1500 m will be jointed before a disconnecting shaft (4) is located to give access to
the directly buried GIL and to hold the disconnecting unit for separating gas compartments of
the GIL. These disconnecting shafts are made of concrete.

Figure 3.7.16 shows how a section of GIL is laid into the trench.

After the GIL is jointed, welded and protected against corrosion, the backfill of the trench
starts. Basically, the soil which has been dug out will be backfilled. Only under certain thermal
conditions and directly at the GIL will a special backfill material be used to increase the heat
transfer conductivity. The GIL is fixed in the ground by friction caused by the weight of the
soil above. The concrete underground shafts are accessible for high-voltage testing and for
monitoring devices for temperature, gas pressure or partial discharge measurements.

The disconnecting units are placed in underground shafts. In these shafts the secondary
systems such as gas density monitoring and arc location systems are also placed [102].

The GIL sections are held by cranes close to the trench and then laid into the trench. The
trench is prepared with a stone-free laying bed and after laying, the trench will be closed by
backfill material. The moment of laying the GIL into the trench is shown in Figure 3.7.17(a),
while Figure 3.7.17(b) shows the bended tube and backfilling of the trench [79].

For the single-phase GIL, the trench will be of about 2–2.5 m depth so that the minimum
soil covering the GIL is 1 m. Shafts are located at distances of approximately 1 km along
the transmission lines. The shafts are made of concrete and hold the disconnecting unit for
the separation of gas compartments. Connected to the disconnecting unit are the gas density
sensor and UHF sensor for partial discharge measurement and for the arc location system.
These concrete underground constructions have to be integrated into the landscape.

In Figure 3.7.18 a trench is shown for two three-phase GIL systems at the moment where two
phases are laid. The trench is opened once, which allows large machines for soil excavation and
parallel laying of single-phase GIL. Finally, the complete trench will be closed with backfill
at once.

Figure 3.7.16 Laying the GIL into the trench. Reproduced by permission of © Siemens AG

For directional changes, angle units are used which are directly buried in the ground. The buried angle unit is made of moulded aluminium and can be made for any angle. The angle unit shown in Figure 3.7.19 has an angle of 30°. The angle unit needs to take the thermal expansion forces which come from the pipe and lead the forces into the surrounding soil.

Figure 3.7.20 shows a section of a GIL between two shafts (1). The underground shafts house the disconnecting and compensator unit (2). The distance between the shafts is between 1 and 1.5 km. Each phase represents one single gas compartment. Also, in the middle of the graphic, the directly buried angle unit (3) is shown as an example. The angle unit is fixed in the soil, and part of the gas compartment of the section [141].

The top view shows three single-phase insulated GIL forming one three-phase electrical system. The distance between the enclosure pipes of the GIL is between 0.5 m and 0.8 m

Figure 3.7.17 Laying the GIL into the soil: (a) laying by cranes; (b) backfill. Reproduced by permission of © Siemens AG

Figure 3.7.18 Directly buried trench for two GIL systems. Reproduced by permission of © Siemens AG and © Amprion GmbH

depending on thermal requirements or on accessibility for repair. With an enclosure pipe diameter of 0.5 m, the total width is 2.5 m to 3.2 m for the three-phase GIL and – including the required space beside the GIL – a total of about 4.5 m to 5 m can be expected for one three-phase GIL directly buried to transmit 2000 MVA.

The trench requirements are related to thermal and mechanical impact on the GIL. The backfill material needs to have defined thermal conductivity for heat transportation from the GIL enclosure pipe into the surrounding soil. The backfill material has a typical content of sand and clay in a mixture to keep water in the soil for heat conductivity. There are physical heat conductivity data given in data sheets for soil. On the mechanical side, the soil needs to be free of stones or rocks in the backfill material to avoid damage to the GIL enclosure pipe and its passive corrosion protection. Usually, the backfill material can be used directly, only some filtering of rocks and stones is necessary. In some cases, for improvement of thermal conductivity, special materials are added. In very rare cases, e.g. pure rock, the total excavation material is replaced by proper backfill material.

In cases when it can be expected that soil drilling may harm the buried GIL, it is necessary to cover the pipes. A sufficient cover of soil usually ensures protection against external damage, e.g. due to agricultural use. A concrete cover plate can be added to protect against drilling or digging in the soil. Huge trees with large roots should be avoided directly above the GIL [79].

Figure 3.7.19   Angle unit directly buried, 30° angle. Reproduced by permission of © Siemens AG

In Figure 3.7.21 the cross-section of a buried GIL is shown. Basically, there are three areas around the GIL. First the bedding material which shall be free of large stones, second the backfill material which is usually the excavated soil, and at a distance the natural soil.

The side view shows the shafts for the disconnecting units, including control and monitoring equipment (gas density sensor, PD sensor and arc location sensor). The shaft needs to be accessible, with an inner height of about 2 m. The length may be 5 m and the width may be 6 m. The shaft can be closed by a plate on the manhole for access.

3.7.5.3   Corrosion Protection

3.7.5.3.1   General

For directly buried GIL, corrosion protection is essential. The water and minerals in the soil would destroy the oxide layer of the aluminium enclosure pipe and then corrode the enclosure. There are two corrosion protection systems available. Passive corrosion protection is a layer of

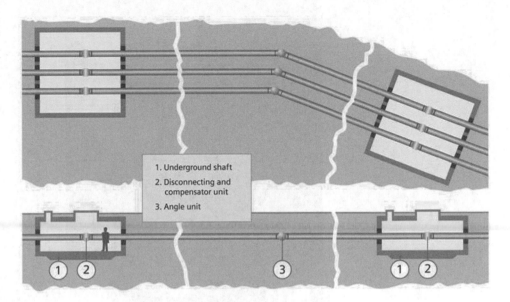

Figure 3.7.20   Three-phase GIL system, top and side view. Reproduced by permission of © Siemens AG

polyethylene or polypropylene including a corrosion protection layer which excludes oxygen from the aluminium. This passive corrosion protection is successfully used for pipelines and gives long-time protection. In addition, an active corrosion protection can be added which protects the aluminium pipe by electrical potential towards a sacrificial electrode. This active corrosion protection may also be added at a later time to the GIL [106, 107, 141].

3.7.5.3.2   Passive Corrosion Protection

The passive corrosion protection is a layer or coating made of polyethylene (PE) or polypropylene (PP) material of 3–5 mm thickness. See also Section 3.3.11.

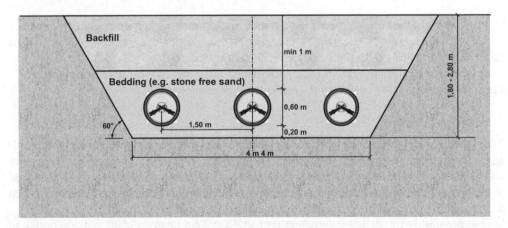

Figure 3.7.21   Cross-section of directly buried GIL. Reproduced by permission of © Siemens AG

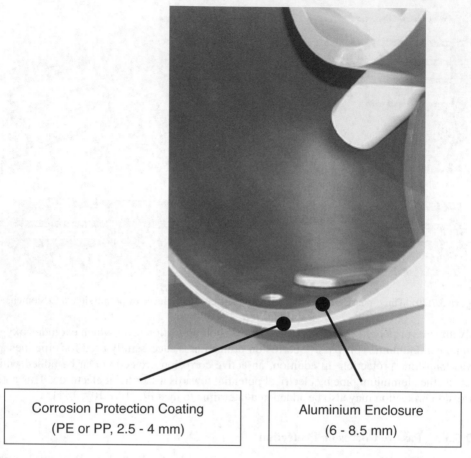

| Corrosion Protection Coating (PE or PP, 2.5 - 4 mm) | Aluminium Enclosure (6 - 8.5 mm) |

Figure 3.7.22   Corrosion protection coating of the enclosure pipe. Reproduced by permission of ©
Siemens AG

The coating of the GIL enclosure pipes is made in the same way as for pipelines. The surface
will be cleaned by acid fluids and/or sand brush and then the coating is extruded to the pipe or
tapes are wound around the pipe with PE or PP coating material.

To protect the jointing weld, shrinkable tubing or spray methods are used. In Figure 3.7.22
the corrosion protection coating of the enclosure pipe is shown.

3.7.5.3.3   Active Corrosion Protection

In addition to the passive corrosion protection, an active corrosion protection system can be
added. The active corrosion protection system adds a voltage potential to the passive protected
and electrically insulated enclosure pipe to protect the aluminium enclosure by the sacrificial
electrode. The voltage is related to the materials and is, in the case of GIL, 1–1.2 V.

One possibility for an active corrosion protection system involves using a diode bridge to
generate 1.2 V (which is twice the gate voltage of 0.6 V for one diode) of protection voltage

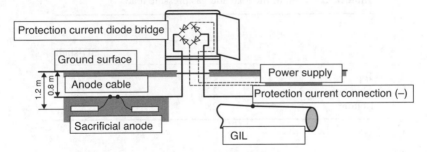

Figure 3.7.23 Active corrosion protection using a diode bridge. Reproduced by permission of ©
Siemens AG

level. The sacrificial electrode is made of iron or graphite and is buried in the soil at a minimum
level of 0.8 m. The negative end of the diode bridge is connected to the buried GIL (see
Figure 3.7.23).

3.7.5.4 Mechanical Stress

3.7.5.4.1 General

The mechanical stresses of a directly buried GIL come from the weight of soil which lays on
top of the pipe. These are relatively high values, as a cubic metre of soil weighs more than one
thousand kilograms.

In addition, in case of traffic the weight of the vehicles must be added, which is done
by wheel loads and depends on the type of traffic (e.g., cars or trucks). These forces are
cross-directional forces to the enclosure pipe.

The thermal forces build up with heat and the absolute temperature of the enclosure pipe
against friction with the surrounding soil, which will avoid movement. These thermal forces
are high, e.g. 180 tons for a 40°C temperature increase.

In Figure 3.7.19 a GIL pipe is shown when laid into the trench surrounded by stone-free
bedding material. In the front and back of the photo the straight section, which is bent with a
bending radius of 400 m, can be seen. In the centre of the photo an angle unit is shown shortly
before being totally buried.

3.7.5.4.2 Cross-Direction Stress

The cross-sectional stresses cover the stress contributions of soil load, traffic loads and gas
pressure load. For the analytical calculations the standard ATV-A127* is used [52, 93].

The calculations are made in two- and three-dimensional models [78].

The calculation model following the ATV rules reflects the method of laying and the type of
soil coverage of the GIL in the trench. The worst condition is the vertical trench wall, where
the stress is the highest (at 26.2 MN/m²). This model takes this highest value to be on the safe

---

*ATV: German Regulation for handling mechanical stresses on pipes with traffic load.

Table 3.7.2 Soil, traffic load and gas pressure load

	σ_{VM} [MN/m²]	
	2D	3D-shell
Top	28	28
Abutment	22	28
Bottom	28	28

side. The deformations in the vertical direction are −3.3 mm and in the horizontal direction 2.9 mm. On this basis it can be said that the chosen stresses are higher than in reality, and therefore some safety margin is provided. This is also shown in the following finite element (FE) calculations.

The FE calculations have been made with two- and three-dimensional models of the GIL. In the model, the soil and the enclosure pipe are defined in 8 or 20 knot square elements. A problem in the model is the current definition of the soil stiffness and deformation ability. This has great influence on the results, and also on the practical laying conditions. According to recommendations of the ATV standard, the stiffness below the enclosure pipe has been chosen ten times higher than above the enclosure pipe [93].

The cross-sectional stresses of the GIL need to be below the maximum allowed stresses as defined in the standard depending on the pipe material and wall thickness. When only soil and traffic loads are taken into account, the maximum values are far below the limits. The calculated values show the tensile force on the outside of the enclosure pipe and the pressure on the inside of the enclosure pipe. The stress on the inside of the enclosure pipe is higher than on the outside.

In Table 3.7.2 the stresses of soil, traffic load and gas pressure are shown. The values indicate the stresses are below the limit of 83.33 MN/m². The maximum stresses are the superposition of soil, traffic load and gas pressure load and are calculated as:

Stress of soil and traffic load: $\sigma_{VM} = 19.3$ MN/m²
Stress of gas pressure load: $\sigma_{VM} = 27.5$ MN/m²
Resulting stress: $\sigma_{VM} = 46.8$ MN/m²

The allowed stress, including a safety factor of 1.5, is $\sigma_{zul} = 83.3$ MN/m². The GIL enclosure pipe can withstand – according to these analyses – the stress from soil, traffic load and gas pressure load.

3.7.5.4.3 Longitudinal Stresses

The model of the GIL has the following dimensions:

- longitudinal (x-axis):   2400 m
- cross-sectional (z-axis):   500 m
- elevation difference (y-axis): 20 m

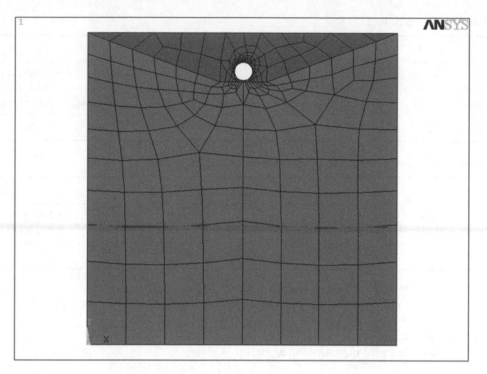

Figure 3.7.24 Model for the mechanical calculation. Reproduced by permission of © Siemens AG

The vertical and horizontal relationship of 20 m/2400 m is less than 1% and can be neglected. The influence of elevation differences is not included in the model because the impact is too small.

A two-dimensional model was set up to calculate the impact of soil, traffic load, gas pressure load and thermal loads of the enclosure of the GIL. The soil coverage on top of the GIL is taken as 2.0 m and the model is shown in Figure 3.7.24. The following laying options have been investigated:

- solid fixed ends of the GIL enclosure pipe,
- free-moving ends of the GIL enclosure pipe.

Traffic loads from streets or railroads need to be added to the gas pressure load, temperature load, soil loads and the own weight of the GIL. Traffic loads from street traffic or railroads are recognized separately because they cannot be at the same location, see Table 3.7.3.

The σ_{VM} of the loads have very different values – from 0.05 MN/m^2 for soil load to 65.8 MN/m^2 for thermal load. The load from the gas pressure and the load from the temperature have opposite vectors.

Free-Moving End of the Enclosure Pipe
Two models have been investigated.
    Case 1: ideal laying of pipe in the soil with equal movement.

Table 3.7.3   Traffic loads

			σ_{VM} [MN/m²]				
Load cases	Road traffic	Railroad traffic	Inner pressure	Temperature	Own weight + soil load	Total (no traffic load)	Total (including traffic load)
Pipe	2.5	6.4	23.7	65.8	0.05	73.9	78.5

Table 3.7.4   Load of traffic, gas pressure, temperature, weight and soil

			σ_{VM} [MN/m²]				
Load cases	Road traffic	Railroad traffic	Inner pressure	Temperature	Own weight + soil load	Total (no traffic load)	Total (including traffic load)
Pipe	2.5	6.4	23.3	20.9	0.4	32.0	37.6

The absolute values of the different loads shown in Table 3.7.4 are superposed to find the total load.

Displacement in x-axis

For a 20 m-long pipe the following expansion is given:

$$\varepsilon = \Delta l / l = \alpha_T \Delta T \rightarrow \Delta l = (2.35 \times 10^{-5})°\text{C}^{-1} \times 40°\text{C} \times 20\,\text{m} = 1.88 \times 10^{-2}\,\text{m}$$

In the FE calculation the horizontal displacement is 1.4 cm, which is 75% of the analytical calculation of the free-moving enclosure pipe.

Friction of Pipe Forward Soil

The connection between soil and enclosure pipe is defined by friction of the pipe forward soil. Therefore, relative displacements between pipe and soil can be found. For a 20 m-long pipe the horizontal displacement is 1.66 cm, which is 88% of the analytical calculation result. To calculate the friction between the pipe and the soil, the following values need to be specified: friction coefficient and contact stiffness between pipe and soil. These values can only be estimated. In this laying method the stress values are lower.

Bent GIL

Different from the vertical displacement of 20 m, the differences in cross-section of 500 m cannot be neglected. To estimate this impact a plate model has been developed where the GIL enclosure sections are fixed in the soil. The model is 400 m long and has a bending radius of 500 m. At each end an angle element is added. For the calculations of the stresses in cross-section, only the contribution of gas pressure and temperature is important, because all other stresses show a vertical vector.

Bending in the y direction has no impact on the x- and z-axes for fixed pipes and movable pipes.

The total stress in the x–z-axis from gas pressure and temperature load is 73.3 MN/m² when the GIL enclosure pipe is fixed and 73.0 MN/m² when the enclosure pipe ends can move freely. This shows that the stresses in cross-sections do not lead to a stress reduction at the free-moving ends because the enclosure pipe is completely enclosed in the soil.

Calculation of Angle Element
Because of the thermal expansion and the gas pressure in the GIL angle element, the angle area sees forces vertical to the pipe axis. Therefore, higher withstand of the bending in the soil is needed in the angle area. To increase the soil volume per metre of GIL pipe in the angle area, the angle unit consists of two pipes with low wall thickness and stronger wall tightness in the centre with a spherical shape. The GIL straight pipes at the two ends are connected to the angle element.

The loads of the angle element consist of the inner gas pressure, the aerial thermal expansion forces (typically 1.2 MN) and the soil load of the covering backfill of up to 2 m. The axial forces give symmetrical loads to the angle element. For the model, the pipe was segmented into 63 elements of the spherical part and the surrounding soil and cross-linked by 4 knots per volume element. The soil has been modelled as elastic.

As an example, only one case of the load of the angle element is presented here. For real applications any of the possible operational conditions related to soil, ambient and conductor temperature needs to be modelled to exclude overstress situations. For the calculation, no friction between the angle element and soil was allowed. This means that the pipe and soil are fixed coupled.

The chosen, strong conditions for this case do not show overstress in the area between the pipe and the spherical part of the angle element. For maximum forces which reach the plastic deformation of the material, a non-linear model needs to be chosen to show the deformation of the angle element. The real situation of the angle element in the soil will have some friction which was excluded from the calculation of the model and, therefore, the forces will be reduced in practice.

Conclusion
The stress analysis in axial direction shows that the thermal and pressure stresses dominate. Together with the loads from soil pressure, the own weight and traffic loads to the GIL, a maximum stress of 78.5 MN/m^2 has been calculated which is less than the allowed 83.33 MN/m^2. The stresses at the angle element need additional calculations to cover the forces which are vertical to the GIL axis.

The additional load of the GIL coming from the elastic bending of 400 m bending radius is low and can be neglected in this calculation.

3.7.5.5   Movement

3.7.5.5.1   Movement of the Enclosure Pipe

The stresses in the pipe are strongly related to the movement of the pipe. No movement leads to a maximum of stress. When movement of the pipe occurs, the stress will be reduced. Calculations with fixed pipes lead to high stresses. The movement of the pipe is related to the friction of the pipe surface with the surrounding soil.

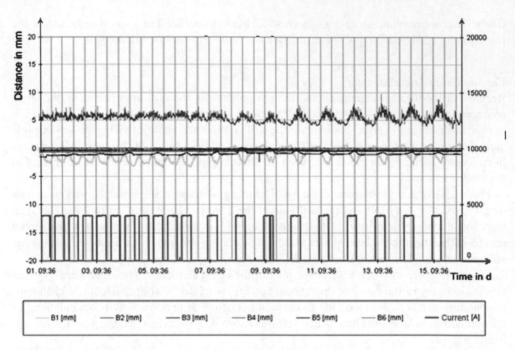

Figure 3.7.25 Movement of the enclosure pipe (upper lines) related to current flow. Reproduced by permission of © Siemens AG

Measurements of movements of the pipe have been made on a 50 m section of GIL. In Figure 3.7.25 the movement of the GIL (upper lines) is shown related to the current flow (lower line) which is heating the GIL.

At different measuring points (Pt 2, 3, 4, 5) along the GIL pipe, the values of Table 3.7.5 show only small movements of 0.5 mm and less.

It is clear that the static friction of the pipe in the soil plays an important part and thereby reduces the expansion of the pipe by 62%. To enable a comparison to be made with the theoretical calculations for long-term testing of the buried GIL, calculations using the FEM were made which ignored friction between the pipe and the soil and included a temperature difference of 40°C.

Table 3.7.5 Movement of the GIL pipe in the soil

	Movements (mm)	Absolute distance (mm)
Pt 2	−0.6/−0.4	0.2
Pt 3	−0.5/−0.3	0.2
Pt 4	−0.1/0	0.1
Pt 5	−1.1/−0.6	0.5

3.7.5.6 Thermal Stress

The thermal stress of a directly buried GIL is related to the current flow in the conductor and the heat transfer into the soil. The thermal design of a GIL is a basic engineering study which needs to be carried out for each project because of changing environmental conditions of the surrounding soil. The internal heat transfer from the conductor to the enclosure through convection of the insulating gas is stable at about 10°C temperature difference. The soil conductivities are very different depending on soil mixtures and the availability of water and moisture.

In Section 3.2.16 the theory for thermal design of GIL is explained.

The low electrical resistance of the GIL has a low thermal heat-up during operation in consequence. When directly buried, the limiting temperatures are given by the non-reversal soil dry-out which requires a maximum long-term temperature increase of below 30°C. The heat transfer and the soil thermal conductivity play a major role in the thermal layout. Because of the large volume of soil around the GIL, some days of higher power transmission and overload currents may be acceptable. The limitations given here are the maximum internal temperatures for conductors (120°C) and insulators (105°C).

The GIL and its surrounding soil form a system of thermally coupled bodies with inner heat production by circulating electric current in both the conductor and the enclosure. Convection and radiation transport the heat losses from the conductor to the enclosure, while heat transfer in the gas gap between the conductor and the enclosure pipe by conduction is negligible. The produced heat from the enclosure dissipates in the soil, mainly in the radial direction, and will end at the surface of the soil. The ambient air at the end will take away the produced heat by convection [142].

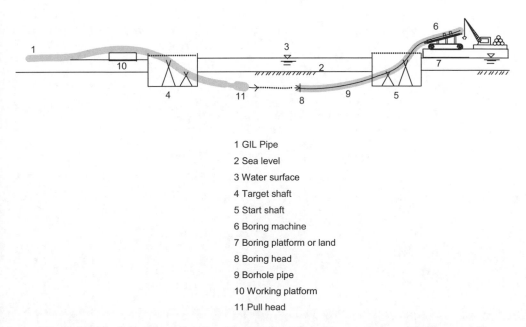

1 GIL Pipe

2 Sea level

3 Water surface

4 Target shaft

5 Start shaft

6 Boring machine

7 Boring platform or land

8 Boring head

9 Borhole pipe

10 Working platform

11 Pull head

Figure 3.7.26 Principle of directional boring. Reproduced by permission of © Siemens AG

3.7.6   Directional Boring

Directional boring, also called micro-tunnelling, is usually used to bypass obstacles like streets, rivers and railroads, which cannot be derouted for the time of the GIL erection. Typical lengths for such micro-tunnels are some hundred metres, with possible lengths up to 1 km. The diameter of a micro-tunnel is between 0.5 m and 1.0 m, and single-phase cables or GIL systems are pulled in [118].

In Figure 3.7.26 the principle of horizontal boring is shown. In this case, the application is to underpass water at sea or in a lake. From the right side the boring machine makes a hole and from the left side the GIL is inserted.

At the start shaft (5) the boring begins with the boring machine (6) placed at a platform on land (7). The boring head (8) moves towards the target shaft (4). Once the boring machine is finished, the hole is secured by a borehole pipe (9) usually made of PE or PP. Through this borehole pipe the GIL is pulled.

The borehole can take single-phase GIL pipes of diameter 0.5 m when the borehole pipe has a diameter of 0.5 m. To pull three single-phase GIL pipes into one borehole pipe the diameter of the borehole needs to be 1.8 m.

3.8

Long-Duration Testing

3.8.1 General

The second-generation GIL has been tested in long-term duration tests with combined stresses of electrical current and high-voltage test voltages. The values for the currents and voltages have been chosen higher than the nominal ratings to prove, under accelerated ageing conditions, the lifetime of the GIL. At the IPH test laboratory in Berlin, Germany, long-term tests have been carried out on tunnel-laid and directly buried GIL in cooperation with leading transmission system operators in Germany.

The use of N_2/SF_6 gas mixtures and the completely welded design of the GIL are new and have been proven by these long-duration tests. The long-duration test gave the proof that the GIL can be laid under realisitic on-site conditions and will function reliably in operation for the expected lifetime of 50 years.

The GIL for the long-duration tests was installed in a tunnel or directly buried in two different test set-ups. For the long-duration test only one single phase of GIL was installed, because the two remaining phases behave the same. The total length of the installations was chosen to allow all functional units to be installed and tested.

The long-duration test included also the proof of the repair process on-site. To replace a repair section of the GIL under realistic on-site conditions gave the proof for the repair process and repair time. Even if such a failure is extremely rare based on the experiences with GIL over the last 35 years, the repair process must be available.

The tunnel-laid GIL test set-up was approximately 70 m long and was installed in a concrete tunnel of the IPH test field. Under real on-site laying conditions the computer-controlled orbital welding technique was applied to prove its practicability. The whole assembling process of pre-assembling the GIL units on-site, the connection of GIL sections, the pulling into the tunnel, the high-voltage conditioning and the test operation sequences for the simulation of 50 years of lifetime has been applied to the GIL test set-up.

The directly buried gas-insulated transmission line was approximately 100 m long and was laid directly into the trench at the IPH test field. The long-duration test set-up included all types of GIL units, has been pre-assembled on-site and was laid into the trench before burying. The outside works under realistic conditions next to the trench and in the trench have been proven,

Gas-Insulated Transmission Lines (GIL), First Edition. Hermann Koch.
© 2012 John Wiley & Sons, Ltd. Published 2012 by John Wiley & Sons, Ltd.

including the automated welding process and the on-site high-voltage commissioning. After that, the long-duration test proved 50 years of simulated lifetime of the GIL.

All test programmes have been successfully completed, with the long-duration tests for GIL with N_2/SF_6 gas mixtures laid in a tunnel or directly buried. The test results, the test set-up and the measurement methods are explained [38, 102, 141, 143–145].

In general, the test results of the long-duration tests showed that GIL with N_2/SF_6 gas mixture is a reliable technology with the ability to bring high power ratings from overhead lines directly underground in one three-phase system. It was shown that for high power transmission ratings of 2000 MVA or 3000 MVA, three-phase systems can be built reliably, economically and ecologically.

Not included in the N_2/SF_6 test set-ups are any types of switching devices like circuit breakers, disconnector or earthing switches. These switching devices are considered as GIS and 100% SF_6 is used. The long-duration test set-ups were equipped with several measuring devices to measure temperatures, movements, mechanical forces, currents and voltages. The measuring results are explained [146].

3.8.2   Tunnel Version

3.8.2.1   Test Set-up in a Tunnel

Together with the leading German transmission system operators, a realistic test set-up was built at the test laboratory IPH, Berlin to qualify the GIL for tunnel applications. The test set-up contains all the required components of a GIL and is connected to the high-voltage test equipment by GIS components.

Long-Duration Voltage and Currents
The long-duration test was set to 2500 hours, representing a lifetime of about 50 years for the GIL system by applying an increased voltage to provide accelerated electrical ageing. The mechanical ageing was generated by the heating up of the conductor and enclosure through electric current, and the following cooling down of the conductor and enclosure which then gives mechanical expansion forces to the GIL system and movement of the conductor in the contact sliding system or the enclosure into the expansion below.

The test includes high-current and high-voltage test sequences to simulate heating and cooling sequences of the GIL. One heating sequence lasts 5 hours, with a current of 3200 A to reach the maximum temperature of the system of 60°C (at a current rating of 2200 A). The cooling sequence lasts 7 hours, to cool the GIL conductor and enclosure pipe to the ambient temperature in the tunnel of approximately 25°C.

During the cooling sequence a phase-to-ground voltage of 480 kV (which is double the rated value) was applied to the GIL (rated phase-to-ground voltage 240 kV), and every 480 hours a switching impulse test voltage of 1050 kV was applied to the GIL.

Long-Duration Test Cycles
The total long-duration test cycles for a 50-year lifetime of simulation had a duration of 2500 h. This total test time was split into five cycles of 480 h and one remaining cycle of 96 h. See Figure 3.8.1.

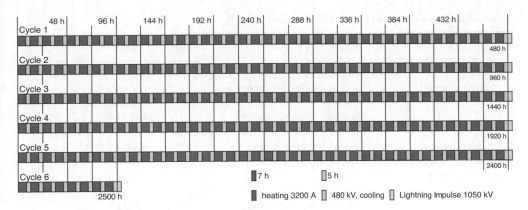

Figure 3.8.1 Long-duration test cycles for heating and cooling sequences and lightning impulse test
Reproduced by permission of © Siemens AG

Each test cycle was split into 12 h sequences with 5 h of heating at 3200 A and 7 h of cooling with no current flow but an applied high voltage of 480 kV. 480 kV phase-to-ground is equivalent to twice the rated voltage and represents the 50-year lifetime simulation of 2500 h. These sequences were repeated until 480 h or 20 days had passed. At the end of each 480 h cycle a lightning impulse voltage of 1050 kV was applied to the GIL to simulate lightning strikes or switching operations of the network [38, 147].

GIL Functional Units
The long-duration test set-up includes all components which are needed to build a long-distance GIL, as follows:

- straight unit
- angle unit
- disconnecting unit
- compensator unit.

The long-duration test set-up is shown in Figure 3.8.2 and the test values are given in Table 3.8.1.

Figure 3.8.2 shows the 70 m-long GIL test set-up for the long-duration test, installed in the tunnel. The tunnel segments with 3 m diameter and 4 m length are laid above ground and connected [6, 36].

The straight unit is the main component used to cover distances. To follow bends, elastic bending can be used. The bending radius can be as small as 400 m. The typical length of the straight unit is 12–18 m and in this case 12 m-long pipes have been chosen. The straight units are jointed by automated orbital welding.

The angle units are used if the bending radius is below 400 m or if a directional change is needed. This is usually at the end of the GIL, when a connection to an overhead line or along the line is needed. The long-duration test set-up has one integrated 20° angle unit close to the left end.

Table 3.8.1 Long-duration test data of tunnel-laid GIL [6]

Type of test	Stresses	Unit	Values
Initial test (commissioning)	Test voltages:		
	Rated short-time AC withstand voltage	Ud	630 kV incl. PD monitoring
	Switching impulse voltage	Us	1050 kV
	Lightning impulse voltage	Up	1300 kV
Long-duration test sequence before repair from 0 to 960 h	5 h-long duration current	I	3200 A
	7 h-long duration AC test voltage	U	480 kV (2.0 p.u. incl. PD monitoring)
	Any 480 h:		
	switching impulse voltage	Us	1050 kV
Interim test after simulated repair (re-commissioning)	Test voltages:		
	Rated short-time AC withstand voltage	Ud	630 kV incl. PD monitoring
	Switching impulse voltage	Us	1050 kV
	Lightning impulse voltage	Up	1300 kV
Long-duration test sequence after repair from 960 to 2500 h	5 h-long duration current	I	3200 A
	7 h-long duration AC test voltage	U	480 kV (2.0 p.u. incl. PD monitoring)
	Any 480 h:		
	switching impulse voltage	Us	1050 kV
Final test after 2500 h	Test voltages:		
	Rated short-time AC withstand voltage	Ud	630 kV incl. PD monitoring
	Switching impulse voltage	Us	1050 kV
	Lightning impulse voltage	Up	1300 kV

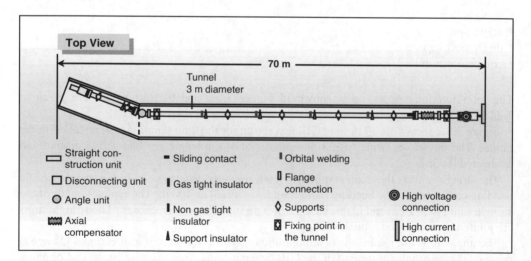

Figure 3.8.2 Long-duration test set-up for tunnel-laid GIL. Reproduced by permission of © Siemens AG

The disconnecting unit is placed at each end of the long-duration test set-up and forms the gas compartment by using gas-tight insulators. In real applications these gas compartments reach a length of 1–1.5 km.

The compensator unit takes the thermal extensions of the GIL enclosure pipe in axial direction and is placed at the right end of the long-duration test set-up before the disconnecting unit. A second compensator unit is placed at the angle unit close to the left end of the test set-up.

In the concrete tunnel the GIL for the long-duration test set-up are fixed to the concrete wall by steel structures. The steel structures are placed at 24 m distances and are equipped with rollers or sliders to allow movement due to the thermal expansion. The tunnel segments were constructed above ground in the IPH test field, as shown in Figure 3.8.3. The concrete tunnel segment has a length of 4 m and an inner diameter of 3 m. The long-duration test set-up has a 50 m section, then a 20° angle and another 20 m section which form a total 70 m length. Such tunnels are usually built deep under soil with a minimum coverage of 10 m, but can be as deep as 40 m under the city in a bored tunnel [36].

In Figure 3.8.4 the tunnel is equipped with two three-phase GIL systems in a total of six enclosure pipes to simulate the real space available in the tunnel. This full tunnel installation was made to have realistic working and space conditions in the tunnel. Only one phase was used for the long-duration tests.

The GIL is fixed on the concrete tunnel wall by steel structures as shown in Figure 3.8.5. The enclosure pipe of the GIL is laid on these steel structures on sliding plates or rollers made of Teflon. The Teflon has graphite addings to make it conductive for grounding of the GIL enclosure pipe solid grounded system. To compensate, thermal expansion longitudinal compensator units are used. These compensators use metallic bellows to compensate the

Figure 3.8.3   Concrete tunnel segments, diameter 3 m. Reproduced by permission of © Siemens AG

Figure 3.8.4   Full tunnel installation with two three-phase GIL systems. Reproduced by permission of © Siemens AG

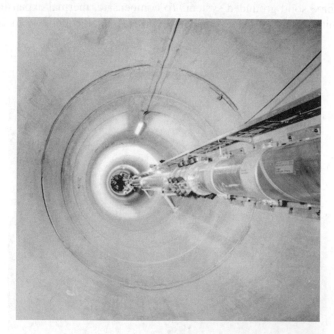

Figure 3.8.5   Long-duration test set-up of one GIL phase in a tunnel. Reproduced by permission of © Siemens AG

thermal expansion of the enclosure pipe when heated up in case of electric current flow (5 hours) or cooling down when the current flow is interrupted (7 hours). The data for the long-duration test programme is given in Table 3.8.1.

3.8.2.2   Test Programme

After the initial commissioning tests the GIL has been applied to a 5 h heating cycle with a current rating 30% above the rated value, so that the maximum temperature of 50°C was reached after 5 h on the surface of the enclosure pipe. During the cooling down sequence an AC test voltage was applied at 480 kV phase-to-ground, which is double the rated value of the phase-to-ground value of the 400 kV system.

In the middle of the 2500 h of total test time, a simulated repair of the GIL was carried out under on-site conditions. The repair included the replacement of a 5 m-long section and recommissioning within only 3 days.

The second sequence of the long-duration test after the repair, as explained in Table 3.8.1, concludes the long-duration test. At the end of the 2500 h-long duration test the same commissioning test with power frequency voltage, switching impulse and lightning impulse voltage was carried out. During the completed long-duration test no problems occurred with the GIL. The PD measurements based on the sensitive UHF measuring method showed constant low values, close to zero. No ageing effect could be detected.

The long-duration test programme as listed in Table 3.8.1 is made visible in the graphic of Figure 3.8.1. The long-duration test starts with type tests of power frequency, switching and lightning impulse voltage according to the GIL standard IEC 62271-204 [54]. This initial type test shows that the GIL test set-up was correctly assembled.

The initial type test was carried out when conditioning the high-voltage system by slowly increasing the test voltage in small steps. In parallel, the partial discharge measuring equipment was connected to the GIL and the PD intensity was permanently controlled. At the end of the initial type test sequence the GIL was free of partial discharges.

The heating and cooling sequences give maximum stress to equipment. When a current of 130% of the rated current is switched on, heating up of the conductor and enclosure will bring maximum expansion of the pipes until the final end temperature is reached after 1–2 h. Then, with switching of the current after 5 h, the conductor and enclosure pipe will shrink and mechanical stress and movement will bring the GIL pipes back to the cold condition. During this shrinking and moving of the GIL a high voltage of double the rated voltage is applied to prove the reliability of the high-voltage system. After several hours (3–4 h) the starting temperature was reached again before (after 7 h) the current is switched on again and the cycle restarts.

This on and off switching of current for heating and cooling and for high-voltage testing during the total time of 105 days simulates the 50-year lifetime of the GIL. Measuring equipment was installed for PD measurement, temperature measurement on the conductor, the enclosure and in the tunnel along the line, movement of the enclosure pipe, the applied voltage and current measurement [34].

3.8.2.3   On-Site Laying in a Tunnel

One important goal of the long-duration test set-up was to prove the practicality of the on-site laying methods. That is where the on-site laying methods have been carried out under realistic working conditions. The good test result proved that on-site laying methods are the right ones.

Special care was taken to prove the cleanliness of the assembly process. It is required for high-voltage gas-insulated systems. In GIS factories this cleanliness is well known and well established in the GIS manufacturing process. The goal was to bring these methods on-site for the assembly of the GIL. The principle is: "If the gas compartment is clean then it will be kept clean until the end of the assembly process". This sounds simple but is not easy to fulfil in a relatively complex assembly process on-site. The excellent testing results of the long-duration test of the GIL proved that the assembly method is correct and reliable.

The sensitive UHF partial discharge measurement method was used to detect the internal clean condition under high voltage. The information about free-moving particles or other inhomogeneous locations in the GIL detected by the UHF sensors is very sensitive and even very small particles are detected. The use of a built-in particle trap proved very reliable, so that the internal gas compartment could be made free of particles during the high-voltage conditioning. As a result, no internal arc was detected during the high-voltage tests.

The GIL has proven its reliability over the simulated lifetime of 50 years and the realistic assembly on-site showed the robust applicability for industrial projects.

Welding

The joints have been welded in a highly automated orbital welding process on-site. In Figure 3.8.6 the connection of the enclosure pipe of the test set-up on-site, inside a welding tent is shown.

Orbital welding of the enclosure pipe has two main challenges: to be gas-tight and to be mechanically solid as a pressure vessel. The gas-tightness does not allow voids in the welds, which can be connected so that gas losses would be possible. The proof is by 100% ultrasonic testing of each weld on-site and the principle of a multilayer weld as shown in Figure 3.8.7.

Figure 3.8.6   Welding of the enclosure pipe. Reproduced by permission of © Siemens AG

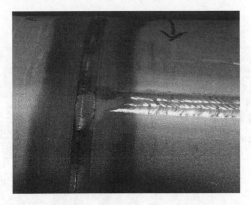

Figure 3.8.7 Multilayer weld of the enclosure in a top view. Reproduced by permission of ©
Siemens AG

This welding method and the ultrasonic weld test were applied under on-site conditions
without any problem, so that this jointing method of aluminium pipes can be seen as a high-
quality and reliable method.

The conductor pipe has different requirements from the enclosure. The conductor pipe weld
needs to be smooth on the surface because it is under high voltage and it needs to carry the
current without high losses.

The electrical aluminium material can be welded by the orbital welding machine with a
surface quality which does not need mechanical rework. This is a great advantage for on-site
work. The welding of the conductor is shown in Figure 3.8.8.

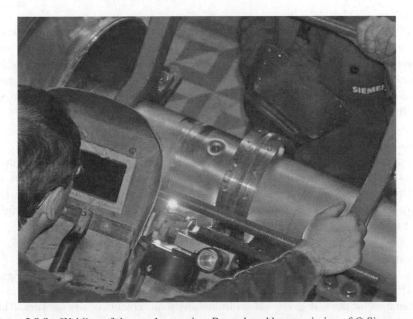

Figure 3.8.8   Welding of the conductor pipe. Reproduced by permission of © Siemens AG

Figure 3.8.9   Gas-mixing device for on-site application. Reproduced by permission of © DILO

Gas Handling
Mixing of the gas has been performed on-site using a newly developed computer-controlled mixing device (see Figure 3.8.9). The mixing process is continuous and has a very high accuracy of the chosen gas mixture in the GIL. The mixed gas is stored in standard gas compartments and is not separated. For a refill, the stored gas mixture is filtered and dried and could be adjusted to gas mixtures with different percentages. The gas-mixing device has a high filling speed, so that long-distance GIL can also be filled in a short time.

Pre-assembly
The pre-assembled GIL segments shown in Figure 3.8.10 are prepared for jointing in the tunnel. The pre-assembly of the GIL segments needs to be done under clean room conditions to avoid dust and particles getting inside the high-voltage gas compartment. Work and surrounding conditions need to be controlled at any time during the on-site laying. A strict on-site quality insurance system needs to be in place to guarantee the clean assembly [36].

High-Voltage On-site Tests
The high-voltage tests of the 400 kV N_2/SF_6 gas mixture GIL follow the recommendations given in IEC 62271-204 [54]. The high-voltage test values for the 20% SF_6/80% N_2 gas mixture are the same as for pure SF_6. The values follow the requirements of the network. These test values are applied during type tests in the factory. On-site, the IEC standard recommends 80%

Figure 3.8.10 Pre-assembled GIL segments. Reproduced by permission of © Siemens AG

of the type test values. The on-site test is no type test and only proves the correct assembly of the GIL. The values are listed in Table 3.8.2 for high-voltage type test values and in Table 3.8.3 for high-voltage on-site test values [75].

Parallel to the on-site power frequency test, partial discharge monitoring is applied [76]. The partial discharge measurement gives clear indication of the internal condition of the high-voltage gas compartment. Using the UHF measuring technology, the high sensitivity gives a clear indication if the on-site assembly of the GIL was made correctly. It is recommended to use this quality control for voltage levels of 400 kV and above [77].

If the GIL is longer than 100 m, then physical effects related to the length have to be taken into account. There is a difference for 10 m (test laboratory set-ups), 100 m (real size prototypes) and 1000 m-long GIL (typical length of a long-distance GIL installation).

At 100 m length the 1.2/50 μs lightning impulse voltage causes reflexions at the open end after 0.3 μs. These reflexions can cause higher voltage levels at different locations of the

Table 3.8.2 High-voltage type test values for 400 kV

Type test	Unit	Value	Condition
Short-time AC withstand voltage (U_d)	kV	630	1 min
Switching impulse voltage (U_s)	kV	1050	15 pulses (50/250 μs)
Lightning impulse voltage (U_p)	kV	1425	15 pulses (5/50 μs)

Table 3.8.3 High-voltage on-site test values for 400 kV

On-site test conditioning	Unit	Value increasing in kV steps	Condition
On-site AC withstand voltage (80%) (80% of short-time AC withstand voltage U_d)	kV	504	1 min

test set-up. The reduced voltage increase steepness can help to limit the maximum values, including the reflexions below the allowed limiting value of $\pm 3\%$. At lengths of 100 m the voltage increase time will be below 20 µs [37, 146].

Impulse voltages generated outside do have a resonance behaviour because such resonance generators are smaller and have lower cost. The resonance test voltage generator uses the capacitance of the test set-up to build a resonance circuit by using inductances. Doubling of the resonance voltage is a typical value. The voltage increase time can be influenced by the choice of the inductive coils. This resonance test equipment is relatively small, and on-site tests can also be carried out in those cases where (formerly) the test equipment was too large to be transported.

Today, variable test voltage frequencies of 30 Hz to 200 Hz are used and will be applied for rated equipment up to 800 kV. The maximum changing current of 10 A limits the maximum size and length of the equipment. For GIL this means that with the equipment available today, on-site tests of 20 km-long GIL are possible.

Partial discharge measurements based on the UHF method have proven to be a very sensitive measuring system for detecting possible internal failures as moving particles or inhomogeneous peaks at the conductor. The UHF partial discharge sensor has a sensitivity to reach up to 500 m into the GIL, to find disallowed particles or inhomogeneous locations. The long-duration prototype is equipped with one sensor at each end of a 100 m-long test set-up. This will take care of the required sensibility.

UHF Partial Discharge Detection
The requirements of the partial discharge sensors (sensibility) have been described in several publications [112, 214, 219]. To prove the validity of the UHF partial discharge measuring system for gas mixtures of N_2/SF_6, a defined needle has been placed at the end of the GIL. With the recording of amplitude, phase angle and impulse number counting it can be verified that interpretations of pure SF_6 defects can be used [207, 208].

Conditioning
When high voltage is applied to the GIL, partial discharges can be detected and visualized by the partial discharge system in the form of diagrams. When the voltage is kept constant, it can be seen that the partial discharge activities are being reduced and will disappear completely after the free-moving particles are kept inside the built-in particle trap. The particle trap is a low electric field area, which will keep free-moving particles away from the high-voltage fields. See Figure 3.8.11 for the so-called APH diagram, which shows the amplitudes, phase angle and number of impulses.

The two detected noise signals are external noises because, in a three-phase system, internal noises are suppressed by the partial discharge measuring system (see Figure 3.8.11).

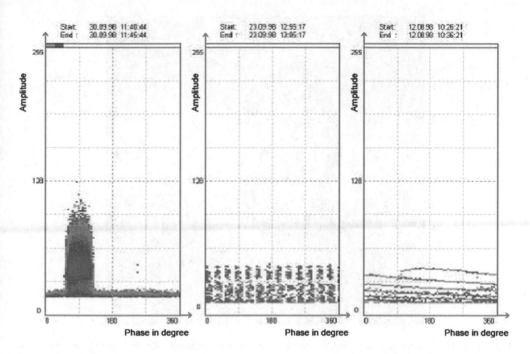

Figure 3.8.11   Partial discharge measurement. Left: defined needle at the conductor, 70 m away from the sensor; middle and right: detected noise signal A and B. Reproduced by permission of © Siemens AG

3.8.2.4   On-Site Repair in a Tunnel

The principle of the repair process is to cut out a section of GIL of some metres length and to replace it with a new GIL unit. Before work is started, the insulating gas needs to be evacuated and stored. Then the enclosure pipe is cut and a gap is opened by pushing the enclosure pipe into the compensator unit. The conductor is cut and the section can be taken out. The repair section is built in just the opposite way, and finally welded by the orbital welding machine.

In Figure 3.8.12 the section of the GIL is shown where enclosure and conductor pipes are cut.

One important part of the long-duration test was to prove the on-site repair under realistic conditions. For this purpose, the single-phase test set-up was extended by five more GIL enclosure pipes to simulate two three-phase GIL systems in the 3 m tunnel. In this area a simulated repair was carried out by cutting out a 2 m length of the GIL and reassembling it with a new GIL section.

This total repair only needed three working days to bring the GIL back to operation. This repair showed that a GIL can be repaired if necessary, in the tunnel and within a short time [77].

Any technique may fail, even if this is a very seldom event – a repair method which can be applied under realistic on-site conditions is necessary. This on-site repair process has been tested under real on-site working conditions after about half the time of the 2500 h-long duration test. A section of the GIL has been cut out and replaced by a repair part. The total repair time was 3 days and the GIL was back in operation [37].

Figure 3.8.12   Enclosure and conductor pipe cut for repair replacement. Reproduced by permission of © Siemens AG

3.8.2.5   Test Results in a Tunnel

The long-duration test over 2500 h and at higher voltage (2 times U_r) and higher current (1.3 times I_r) proved the reliability of the tunnel-laid GIL. This test includes a realistic assembly process on-site and a repair process after about half the total time. All the single-process steps – like welding, gas handling, assembly, cleaning, conditioning and site testing – were applied without any problem. The long-duration test represents a 50-year lifetime of the GIL and proves high reliability.

Practically No Ageing
The long-duration test showed that there is practically no ageing detected. The complete set of high-voltage test sequences was applied to the GIL at the end of the 2500 h-long time test without any failure. Power frequency voltage, switching impulse voltage and lightning impulse voltage have been applied to the GIL, which has been tested for 2500 h at 130% of the rated current [195].

Reliable N_2/SF_6 Gas Mixture
GIL with N_2/SF_6 gas mixtures has proven its high reliability in this long-duration test. N_2/SF_6 gas mixtures have the principle effect of high electronegativity and with this the corona-stabilizing factor starts to dominate at SF_6 percentages below 10%. When the SF_6 proportion is 20%, almost 80% of the insulating capability of pure SF_6 is reached. The

arc-distinguishing capability of SF_6 is strongly reduced by more than 50%, but is not needed in a GIL without any switching device.

Conclusion
The prototype for type tests with a length of about 30 m and the tunnel-laid long-duration test with a length of about 70 m showed positive results of all tests carried out. Electrical, thermal and mechanical requirements have been fulfilled in laboratory tests and under realistic on-site conditions with the long-term test set-up.

Proven test sequences from the 40 years of GIS experiences have been adapted for GIL needs. For high-voltage testing the resonance generator test equipment with variable frequencies and UHF partial discharge monitors is recommended [75].

3.8.3   Directly Buried Version

3.8.3.1   Test Set-up Directly Buried

At the IPH test laboratory in Berlin, Germany, a 100 m-long GIL was tested in a long-duration test of 2500 h. The GIL test set-up was directly laid into the soil between two underground concrete shafts. The on-site assembly works and the laying process were done under realistic construction site conditions. The GIL test set-up included all the GIL units – straight units and angle units – which are directly buried, and the disconnecting unit and compensator unit which were placed in the two shafts at the end. In Figure 3.8.13 the test set-up is shown. The test set-up also includes a "space curve" for the straight GIL directly buried. This "space curve" is

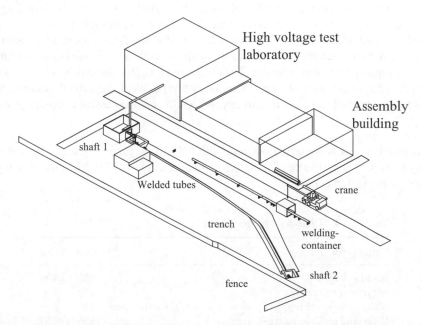

Figure 3.8.13   Site arrangements of the directly buried long-duration test. Reproduced by permission of © Siemens AG

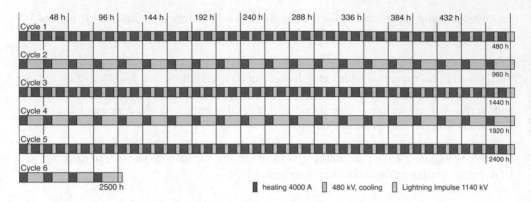

Figure 3.8.14   Long-duration test cycle. Reproduced by permission of © Siemens AG

bent into a horizontal and vertical direction to follow a bending radius of 400 m. The increased test voltage of 480 kV initiates electrical ageing to simulate an electrical lifetime of 50 years. The total test programme has six cycles of 480 h each, as shown in Figure 3.8.14.

Long-Duration Test on a Directly Buried GIL
The long-duration test for the buried GIL has been carried out on a 100 m-long test set-up shown in Figure 3.8.13. Figure 3.8.13 shows the IPH high-voltage test laboratory with the high-voltage connection to shaft 1.

From shaft 1 the trench with the directly buried GIL, of 100 m length including elastic bending and a directly buried angle module, proceeds to shaft 2 at the end. In shaft 2 a ground switch closes the current loop. The current injection devices are in shaft 1.

In a welding container situated beside the trench, the straight GIL segments are jointed by using an orbital welding machine. A crane transports the assembled GIL unit from the nearby assembly building to the welding container. The electrical part of the directly buried GIL is equivalent to that of the tunnel-laid GIL. Therefore, assembly, gas handling, orbital welding and testing are similar. A special feature of the directly buried GIL was the passive corrosion protection.

Parameters
The long-duration test was carried out on a 400 kV GIL system which has been designed following the IEC type test values for dielectric and current tests [54]. The dielectric type test values are shown in Table 3.8.4.

Table 3.8.4   Dielectric type test values

Voltage type test	Unit	Value
Rated voltage U_r	kV	400
Maximum voltage U_m	kV	420
Short-time power frequency voltage with PD measurement	kV	630
Switching impulse voltage	kV	1050
Lightning impulse voltage	kV	1425

Table 3.8.5   Short-circuit withstand test values

Current type test	Unit	Value
Short-circuit peak current	kA	165
Short-time current	kA	63
Duration of short-circuit current	s	3
Arc fault current	kA	63
Duration of arc fault current	s	0.5

All the test set-ups passed the dielectric type tests. Partial discharge measurements have been carried out together with the power frequency test voltages using the UHF partial discharge monitoring system. This measuring technique allows very sensitive measurements inside of the gas-insulated system to detect small disturbances even over long distances. The coaxial GIL system is an ideal high-frequency transmission system, so that even small intensities of approximately 5 pC are detectable over long distances.

The UHF partial discharge measuring technique has been proven as a very effective tool for the conditioning process of the GIL under high voltage. The short-circuit withstand test values are shown in Table 3.8.5.

The current tests have been carried out on test set-ups which implement all GIL units: the disconnect unit, the straight unit, the angle unit and the compensator unit. All the tests have been successfully completed without any problems.

Long-Duration Test Cycles
The laying and commissioning process was checked for practical and suitable laying processes with a 100 m-long test set-up to simulate 50 years of service lifetime. During this long-duration test over a total of 2500 h, electrical and mechanical stresses have been applied to the GIL by heating current of 400 A for 8 h. After one heating sequence a cooling sequence was applied, of 4 h and 16 h continuously together with an applied high voltage of 480 kV phase-to-ground, which is twice the rated voltage. This higher voltage accelerates the ageing speed so that in about 3 months the electrical lifetime of 50 years is concentrated.

Each sequence of 480 h or 20 days is alternately repeated with short and long cycle sequences until the total time of 2500 h is met. At the end of each cycle the switching impulse voltage of 1140 kV was applied to simulate switching operation in the network.

3.8.3.2   Test Programme Directly Buried

Additional typical elements of secondary technology for monitoring and operation have been integrated in the long-duration test. Partial discharge measurements are used for quality control of the high-voltage insulation during the commissioning process and during the complete test time of 2500 h. Gas density of the insulating gas is permanently measured by measuring the gas pressure and gas temperature and calculating the gas density. An electronic sensor for surge impedance travel waves as they occur is integrated to detect the location of an internal arc.

Along the GIL, temperature sensors are placed at top, bottom and side positions of the enclosure pipe. The temperature of the conductor pipe is measured by radio frequency connected surface wave filters to give information about the actual temperature of the conductor. At the ends and in the middle, distance sensors detect the movements of the enclosure pipe when electric current is heating up the GIL.

Measured Data

During long-duration testing, the major physical data is measured to get information for thermal, mechanical and electrical calculation tools. These calculation tools are later used in project planning to define the layout of the GIL.

The single-phase, long-duration test of the directly buried GIL gives realistic proof of the functionality of the GIL. As a transmission technology for high power ratings, the GIL is a future underground long-distance transmission system as an alternative or supplement to today's use of overhead lines [46, 74, 141, 142, 146]. The GIL conductor and enclosure temperature as well as GIL movement due to thermal expansion/contraction were monitored during load cycles. All tests were performed successfully [147].

Comparison of Directly Buried and Tunnel-Laid Test

The requirements and dimensioning of the tunnel-laid GIL and the directly buried GIL are different. Because of the different heating and cooling, the cycle times are 12 h for tunnel-laid GIL and 24 h for directly buried GIL. The current rating for the tunnel-laid GIL was 3200 A and for the directly buried GIL it was 4000 A. Therefore, the diameter of the tunnel-laid GIL is 500 mm and that of the directly buried GIL is 600 mm.

To check the GIL system's suitability for practical use, every effort was made to implement a test set-up which comes close to real conditions. Therefore, the tunnel-laid GIL was assembled on-site and installed in a tunnel made of concrete tubes (total length: 70 m). The directly buried GIL was laid in soil accordingly (total length: 100 m). The pre-qualification tests consisted of several parts. See Tables 3.8.6 and 3.8.7.

Table 3.8.6 Comparison of test parameters of tunnel-laid and directly buried GIL

	Test parameters GIL, tunnel-laid		Test parameters GIL, directly buried	
Commissioning	AC withstand test, 1 min	630 kV	AC withstand test, 10 s	550 kV
	Lightning impulse test	1300 kV	AC withstand test, 1 min	504 kV
	Switching impulse test	1050 kV	Lightning impulse test	1140 kV
	PD monitoring		PD monitoring	
Re-commissioning,	AC withstand test, 1 min	630 kV	AC withstand test, 10 s	550 kV
after	Lightning impulse test	1300 kV	AC withstand test, 1 min	504 kV
demonstration of	Switching impulse test	1050 kV	Lightning impulse test	1140 kV
repair process	AC, 48 h	480 kV	PD monitoring	
	PD monitoring			
Final test	AC withstand test, 1 min	630 kV	AC withstand test, 10 s	550 kV
(tunnel-laid, after	Lightning impulse test	1300 kV	AC withstand test, 1 min	504 kV
2500 h)	Switching impulse test	1050 kV	Lightning impulse test	1140 kV
(directly buried,	AC, 48 h	480 kV	PD monitoring	
after 2880 h)	PD monitoring			

Table 3.8.7 Comparison of load cycles for tunnel-laid and directly buried GIL

	Test parameters GIL, tunnel-laid			Test parameters GIL, directly buried	
Load cycles	Total duration	2500 h	Load cycles	Total duration	2880 h
	Duration of one cycle	12 h	(time	Duration of one cycle	12/24 h
	Number of cycles	210	parame-	Number of cycles	120
	Heating current, 7 h	3200 A	ters	Heating current, 8 h	4000 A
	High voltage, 5 h	480 kV	change	High voltage, 4/16 h	480 kV
			every		
			480 h)		
Intermediate tests, every 480 h	Switching impulse test	1050 kV	Intermediate tests, every 480 h	Lightning impulse test	1140 kV

The intention of the long-duration test was to simulate the electrical and mechanical stresses that would occur over a lifetime of 50 years. This was achieved by the application of load current and high-voltage cycles of the test parameters.

3.8.3.3 On-Site Laying Directly Buried

There are two main principles for on-site assembly and laying. One is to connect longer GIL sections right next to the trench (e.g., 200–300 m) and then lay the GIL section into the trench. The other is to connect the GIL sections above, in a shaft to the trench. Such shafts are located typically at 1000 m distance. The GIL sections are then pulled into the open trench. Which method fits best depends on the on-site conditions, site accessibility and usability of machines and tools.

Assembly Tent
Orbital welding, ultrasonic weld test, on-site gas mixture, on-site corrosion protection and on-site testing are developed to meet the conditions which can be expected on-site: humidity, dust, temperature changes, rain, snow, ice and so on. To meet the quality requirements, an assembly tent is used as shown in Figure 3.8.15.

The installation of GIL under high-voltage clean conditions has proven the suitability of this laying process. The usability of the tools and procedures developed for this solution are successfully demonstrated.

Final assembly on-site takes place either beside the trench or in the shaft structures. The place of assembly depends on the civil engineering design dictated by local conditions. The installation finishes with the laying in the final position.

Trench
Constructing the trench and laying the GIL closely follow the pipeline laying technique. Figure 3.8.16 illustrates a typical trench profile for the directly buried GIL. The thermal expansion is absorbed by the surrounding bedding material by means of friction forces. In addition, the bedding dissipates the heat losses. The heat transfer from the GIL enclosure

Figure 3.8.15   Assembly tent. Reproduced by permission of © Siemens AG

to the soil must be ensured by efficient thermal conductivity of the bedding material. The temperature at the transition from the enclosure to the ground shall not exceed 50°C according to IEC standards to avoid permanent soil dry-out. The GIL in the trench shown in Figure 3.8.15 can transmit 2250 MVA continuously [38].

Shaft
The shaft structures accommodate the separating modules and expansion fittings. In addition, the secondary equipment with the telecommunications system is also located there.

In Figure 3.8.17 the principle of shafts is shown in cross-section (upper part) and bird's-eye view (lower part). The principle shows GIL sections of about 1 km in length. This length represents one gas compartment for the high-voltage part. In the shaft, the disconnecting unit (1) separates the gas compartments and gives access to the gas monitoring sensors, the partial discharge sensors and the gas filling valves.

The thermal compensation unit covers the thermal expansion of the GIL in the shaft before the GIL enters the trench as a straight unit (3), shown with a bending radius of 400 m. In the trench, an angle unit (4) for directional change may be buried before the second shaft is

(1) Disconnection unit
(2) Compensator unit
(3) Straight GIL with 400 m bending radius
(4) Angle unit
(5) Shaft

Figure 3.8.16 Trench profile. Reproduced by permission of © Siemens AG

(6) Trench (1–2.9 m in depth)
(7) High voltage and current connection

is reached. The depth of the trench is typically 1–3 m. The shaft may also be used to connect the GIL to an overhead line or to a GIS in a substation.

3.8.3.4 Repair Process Directly Buried

In addition to the above-mentioned test with mechanical and electrical stresses, after 960 h of the long-duration test the sequence was interrupted and a planned repair process (including the substitution of a tube length) has been carried out. See Figure 3.8.18. The total process of exchanging a segment of the GIL, including the recommissioning high-voltage testing, has been finished in less than two weeks.

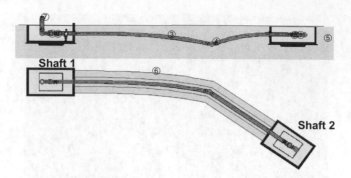

Figure 3.8.17   Shafts. Reproduced by permission of © Siemens AG

Mixing of the gas was performed on-site using a newly developed computer-controlled mixing device. The mixing process is continuous and arrives at a very high accuracy of the chosen gas mixture in the GIL.

The most severe failure of a GIL is an internal arc. The repair process is split into diagnostic and repair time. The diagnostic time is when the failure will be analysed and the damage evaluated using the arc fault location measuring system, the current diagram and the time of the arc.

During this first one to three days all the tools and equipment will be brought to the site location. The repair starts with site installation, opening the trench and disassembling the damaged part of the GIL. This will be on day 4 and 5. The new GIL segment is then fitted in and the GIL will be filled with the insulating gas mixture on day 6 and 7, including the on-site pressure tests. The commissioning will take place on day 8 and 9. With the re-energizing on day 10 the repair is concluded.

The gas mixture can be stored in standard high-pressure gas containers (up to 20 MPa) and subsequently reused after recommissioning. The repair sequence shows that the GIL can be

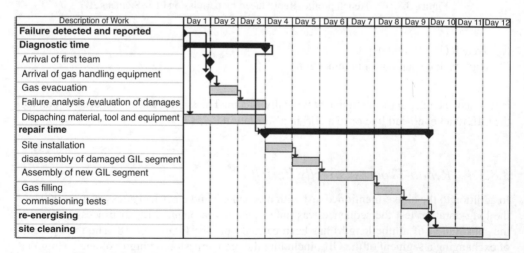

Figure 3.8.18   Repair schedule. Reproduced by permission of © Siemens AG

Table 3.8.8   Tests on commissioning and recommissioning [38]

Test	Measure
Pressure test	Verification per pressure vessel regulation
Gas-tightness test	Checking of flange joints
State of gas mixture	Mixture ratio Filling pressure Dew point
Corrosion protection coating voltage test	10 kV/1 min
Resistance test	Main circuit

repaired on-site without any problems in terms of going back into service. The repair process needs only simple tools and is easy to carry out in a short time.

Commissioning Test

For the purpose of commissioning, comprehensive electrical and mechanical tests are necessary in order to assure the properties of the directly buried GIL. In addition to the verification of the dielectric properties and checking of the secondary equipment used, the following tests must be carried out as shown in Table 3.8.8.

In setting up the long-term test, a three-dimensional curve was simulated to take account of the minimum permissible bending radius of the enclosures. A service life of about 50 years is verified by means of increased voltage and current stress.

Secondary Equipment

In addition, the typical elements of the secondary equipment of the GIL are employed. Thus, PD measurement is carried out during commissioning and online during the test. The gas properties, such as temperature and pressure, are monitored on a continuous basis. Arc location is carried out. The conductor and enclosure temperature and gas density of the GIL are determined by means of radio-controlled sensors at several locations. Displacement sensors in the shaft structures and along the route record the movement of the GIL relative to the ground or to the building, in order to study the mechanical behaviour of the GIL.

During the course of the long-term test, the essential physical variables which describe the GIL – and are used for adjustment of the predictions – are recorded. In addition to the electrical stress imposed on the system by voltage and current, the above-mentioned temperatures and movement are recorded [38].

3.8.3.5   Thermal Calculations Directly Buried

3.8.3.5.1   General Directly Buried

The thermal design of the GIL laid in soil is a complex engineering task, as explained in Section 3.2.16. It is essential for reliable operation of the GIL and to avoid overheating. The long-term test also offers the possibility of proving the models used for thermal calculations.

To confirm the reliability of the GIL, a buried single phase in a long-term test over 6 months was laid in the earth and tested at Berlin test laboratory, at a depth of between 0.7 m and 3 m. The total length of the pipe was approximately 100 m. The buried GIL was cyclically subjected to high current rating (4000 A) and high voltage, which correspond to a heating and cooling phase. The duration of the cycles and the current load were selected such that the temperatures of the enclosing tube and the conductor stayed below the permitted temperatures and at the same time simulated a service period of 50 years. Extensive temperature measurements were taken in order to monitor temperatures in the GIL and also to be able to compare them with calculated values. The GIL was in operation without a break from the beginning of August until the end of November (summer and autumn conditions). The long-term testing of the buried GIL extended to a total of 159 subcycles over 104 days, of which 120 short subcycles occurred in 60 days and 39 long subcycles in 44 days. For the entire duration of test operation the line was subjected during the heating phase to a current of 4000 A.

The GIL and its surrounding soil form a system of thermally coupled bodies with inner heat production by circulating electrical current in both the conductor and the enclosure. The convection and radiation remove the heat losses from the conductor to the enclosure, while the heat transfer in the annulus by conduction is negligible. Then this heat, added to the losses by Joule effect from the enclosure, dissipate in the soil mainly in the radial direction to the surface of the soil – where it flows into the ambient air by the convection. The soil parameters were taken from various literatures on Berlin soil properties.

Before performing the unsteady-state study of the thermal behaviour of GIL, a steady-state model was developed taking into account the mechanisms of conduction in a solid body, the natural convection in a cylindrical cavity, and the radiation and convection in the interface between the soil surface and the air. The thermal system is divided into two parts (GIL and the surrounding soil), looking at the physical phenomena occurring in each. The FEM method (ANSYS program) was used first to check the accuracy of the developed analytical model, then to carry out the unsteady analysis of the thermal behaviour of a buried GIL as presented [33, 49, 78, 79].

The model developed here takes into account the mechanisms of the conduction in a solid body, the natural convection in a cylindrical cavity, and the radiation and convection in the interface between the soil surface and the air. The thermal system is divided into two parts (GIL and the surrounding soil), looking at the physical phenomena occurring in each one.

In Table 3.8.9 the parameters for the calculation are given for the GIL. The power transmission capability is defined by the rated current of 1700 A. At a rated voltage of 420 kV, this is equivalent to a power transmission capability of about 1250 MVA. The thermal resistivity of the soil and backfill surrounding the GIL or the cable is taken as 1.2 Km/W.

Table 3.8.9 GIL technical parameters

GIL	Conductor	Enclosure	Anti-corrosion
Material	Aluminium alloy	Aluminium alloy	PP coating
Outer diameter (mm)	180	517	523
Thickness (mm)	10	8.5	3
Electric conductivity (m/Ω.mm^2)	30.00	21.30	x
Thermal conductivity (W/mK)	x	x	0.286
Radiation emissivity	0.2	0.2	x
Thermal conductivity (W/mK)	x	x	0.286
Current (A)	1700	1700	x

Table 3.8.10   Cross-section parameters

Parameter	Unit	Value
Cover h	m	$h = 0.7$ ($h = 2.6$ for the second location)
Thermal conductivity λ	W/mK	$\lambda = 1.6$
Soil temperature T_s	°C	$T_s = 15$
Initial values for soil temperature T_i	°C	$T_i = 20$

3.8.3.5.2   Calculation Model Directly Buried

Calculations were carried out using the finite elements method. Heat loss, the heat transfer coefficient and the thermal resistance in the annular gap between the conductor and enclosing tube are calculated using a steady-state method according to the IEC 60287 standard [53] and used as constants in the transient calculation. Calculations for the GIL cross-section at location 1 were carried out with the parameters of Table 3.8.10.

The thermal resistance of the soil was measured at the start of the test at three different places (at the ends of the line and in the centre). At each point, measurements were taken at two points at a depth of 0.90–2.30 m. The average thermal resistance measured varied from 0.46 to 0.80 mK/W – in other words, a 70% difference between the extreme measured values. The measurements show a wide scatter from the mean value. The thermal resistance which was used in the calculations was taken as the mean values of the measurements. The boundary conditions are given in Table 3.8.11.

Test Run
The buried GIL was then operated at 4000 A for a sequence of 8 h, and then the current was switched off for 4 h in a short cycle. The long cycle has 4000 A for 8 h and no current for 16 h, see Table 3.8.12. These test cycles were applied for 2500 h in total, with short and long cycles of 480 h each.

3.8.3.5.3   Comparison Calculation/Measurement Directly Buried

Depth: h = *1 m*
In order to compare the measured temperatures with the calculated temperatures, heating of the GIL during the whole test time was simulated. Comparisons of the measured and the calculated temperatures for 16 days (period 1.9.99 to 16.9.99), with two cycles occurring, above (TM1), below (TM3) and to the side (TM3) of the enclosing tube (location 1) are shown in Figure 3.8.19. The calculations agree well with the measured values. The maximum

Table 3.8.11   Boundary conditions

Parameter	Unit	Value	Comment
Interface between soil and air	W/m²K	20	Heat transfer coefficient
Air temperature			Measured air temperature
Temperature of soil	°C	15	20 m away from the GIL
Initial temperature of the soil	°C	20	
Bisecting line	W/m²	0	Heat loss (symmetry conditions)

Table 3.8.12 Cycle load of the buried GIL

Parameter	Unit	Value	Comment
Short cycle			
8 hours/= 4000 A	W/m	145	Loss
4 hours/= 0 A	W/m	0	Loss
Long cycle			
8 hours/= 4000 A	W/m	145	Loss
16 hours/= 0 A	W/m	0	Loss

temperatures rose slowly during the short cycles and after 8 days reached 35°C. During the second period the cooling phase was extended from 4 h to 16 h, which was why the temperatures in the GIL system fell. In this case the maximum enclosing tube temperatures were less than 33°C.

Calculations, unlike measurements, show that the maximum temperature is to be found around the circumference of the underside of the enclosing tube since the effect of heat transfer by natural convection from the inner conductor to the enclosing tube is not taken into account in the calculations. In the upper and lower parts of the enclosing tube the different temperatures are to be explained by variations in the resistance of the soil.

Figure 3.8.20 shows the temperature distribution of the calculations in the soil at time 188 hours (7.8 days), heating phase; the temperature measured during the test is assumed to

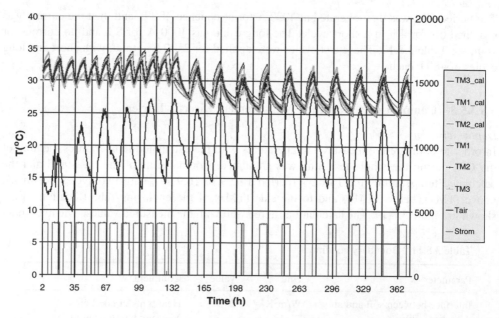

Figure 3.8.19 Measured and calculated temperatures at 1 m depth: comparison of numerical and experimental results of overload current rating of a directly buried GIL; long-duration test, short and long cycle. Reproduced by permission of © Siemens AG

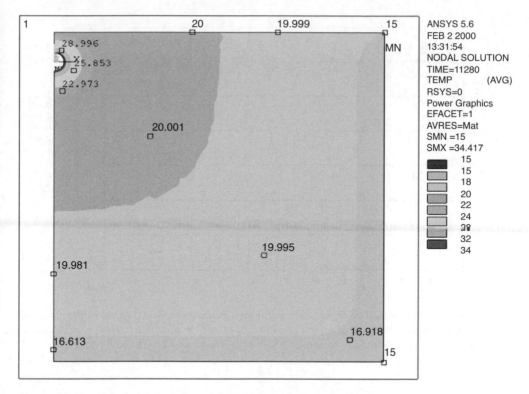

Figure 3.8.20 Calculated temperature after 188 hours at 1 m depth. Reproduced by permission of ©
Siemens AG

be a marginal condition for the air temperature. The temperature distribution during the day
and the night shows a difference only in the higher layer of soil immediately below the ground
surface. This can be explained by the heat transmission between air and the surface of the
ground due to the low air temperature during the night, heat transmission from the GIL is
better since the temperatures in the soil are slightly lower. The fluctuations in air temperature
between night and day do not have such a great influence on the temperature distribution in
the GIL and the soil as those caused by load variation [78].

Depth: h = 2.9 m
A further simulation of the test was carried out for a depth of 2.9 m during the same period
as above (1.09.99 to 16.09.99). A comparison of calculations and measurements is shown in
Figures 3.8.21 and 3.8.22.

Here too there is good agreement between the calculations and the measurements, particularly during the first period when the GIL was subject to current in short cycles. The
calculations show that the temperatures at the bottom and top are higher than the temperature
at the sides, and there is a temperature difference of less than 2°C. The temperature at the
bottom is slightly higher than the temperature at the top ($dT \leq 0.5°C$).

In contrast to this, the test showed a considerable temperature difference between the bottom,
the top and the sides with a high value at the top and a low value at the bottom. In this example

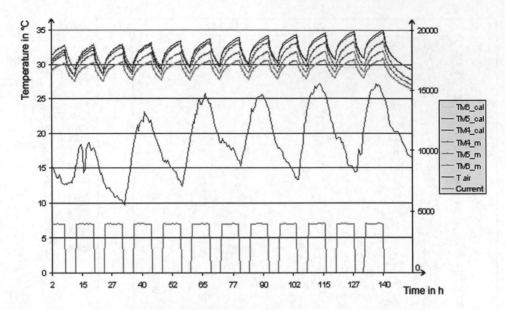

Figure 3.8.21 Comparison of numerical and experimental results of overload current of a directly buried GIL, long-duration test, depth 2.9 m, short cycle (8 h on, 4 h off). Period from 2.09.99 (corresponds to 0 h in the diagram) to 7.09.99 (corresponds to 155 h in the diagram). Reproduced by permission of © Siemens AG

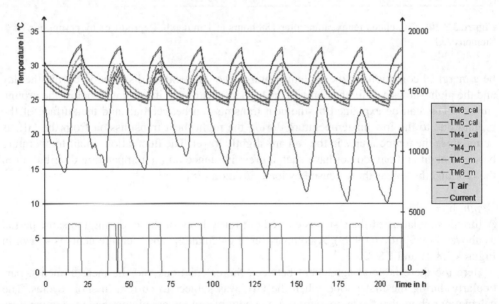

Figure 3.8.22 Comparison of numerical and experimental results of overload current of a directly buried GIL, long-duration test, depth 2.9 m, long cycle (8 h on, 16 h off). Period from 7.09.99 (corresponds to 0 h in the diagram) to 16.09.99 (corresponds to 255 h in the diagram). Reproduced by permission of © Siemens AG

the temperature at the circumference of the pipe is not constant on account of the effect of natural convection between the inner conductor and the enclosing tube not being unsteady.

3.8.3.5.4   Overload Case Directly Buried

In this section the thermal behaviour of a buried GIL and its surrounding soil corresponding to a real application case is presented. A two-GIL system (six phases) is buried underground at a depth of 1.2 m (distance from the soil surface to the tube axial axis) and the axial distance between phases is 1.3 m. The outer surface of the GIL enclosure is protected by a polypropylene anti-corrosion layer with a thickness of 3 mm and thermal resistivity of 3.5 Km/W. The thermal resistivity of the soil and backfill surrounding the GIL is taken as 1.1 Km/W. The soil temperature is taken as 35°C and the air temperature as 40°C. These values are used to calculate the summer ratings of buried systems in warm areas. The insulation gas was taken to be a mixture of nitrogen (N_2) and sulphur hexafluoride (SF_6) at a pressure of 0.7 MPa (absolute) at 20°C.

Results of two types of calculation are presented here, the first is the unsteady state with constant current rating ($I = 1700$ A) and the second is the unsteady state with various cycle loads of current rating (1700 A, 2040 A, 2500 A, 3400 A).

Type 1 Calculation
Unsteady-state calculation with a continuous constant current rating (100% load factor), the simulation time is 4500 h (6.25 months).

The simulation shows (Figure 3.8.23) that after 4500 h (188 days) of service, the fully steady state is not reached and the enclosure temperature is about 60°C. The middle phase is the warmest one, because of the heat input of the neighbouring phase (about 69°C at steady state; Figure 3.8.24). The maximum temperature for each phase is at the bottom and at the side. On the side because of the influence of the other phases, and at the bottom because the thermal resistance from the bottom to the soil surface is greater than from the top to the soil surface.

Calculation parameters:

> Phase 1: Maximum temperature of the warmest phase, at the bottom of the middle one.
>
> Phase 2: Maximum temperature of the second phase from the outside at the bottom.
>
> Phase 3: Maximum temperature at the side of the outside phase.

Type 2 Calculation
Unsteady-state calculation with a load factor of 100%, 120%, 150% and 200%. The simulation was carried out for a rating current of 1700 A with an initial temperature value of the system and the surrounding soil of 35°C and after 2400 h (100 days) the system was overloaded with 120% or 150% or 200% during 720 h (30 days) (see Figure 3.8.25).

- From 0 to 2400 h: continuous current rating 1700 A.
- From 2400 to 2880 h: load cycle 1 includes two periods of 24 h, each of which has a corresponding current rating of 2040 A and 1700 A.

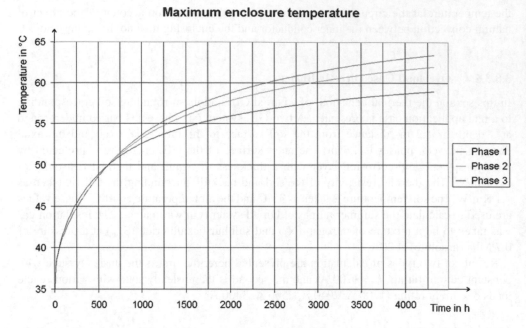

Figure 3.8.23 Continuous constant current rating of 100% load factor after 4500 h. Reproduced by permission of © Siemens AG

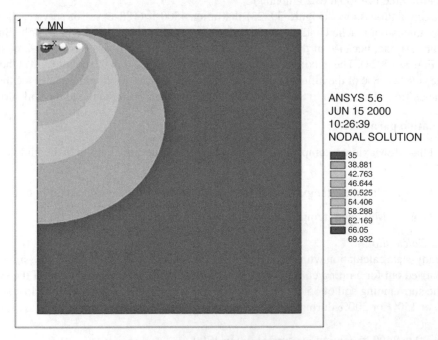

Figure 3.8.24 Continuous constant current rating of 100% load factor at steady state. Reproduced by permission of © Siemens AG

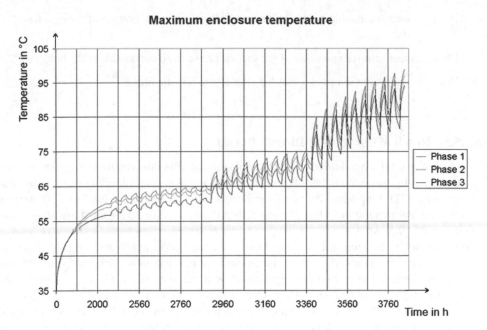

Figure 3.8.25   Overload current. Reproduced by permission of © Siemens AG

- From 2880 to 3360 h: load cycle 2 includes two periods of 24 h, each of which has a corresponding current rating of 2550 A and 1700 A.
- From 3360 to 3840 h: load cycle 3 includes two periods of 24 h, each of which has a corresponding current rating of 3400 A and 1700 A.

The simulation shows that after 100 days of service the fully steady state is not reached, and the enclosure temperature is about 60°C. With an overload of 200% of the GIL, the maximum temperature of the conductor (105°C), respectively 95°C enclosure temperature is reached after 200 h (8.3 days). With an overload of 150% the simulation shows that after 3200 h (133 days) the maximum allowed temperature of GIL is not reached. This means that the GIL can transport at a continuous rating of 2550 A and can be 200% overloaded over 8 days.

Calculation parameters:

Simulation of 2-system of buried GIL with current rating of 1700 A.

Simulation time 3840 h (160 days)

from 0 to 2400 h: 1700 A

from 2400 to 2880 h: load cycle 1 (24 h at 2040 A, 24 h at 1700 A)

from 2880 to 3360 h: load cycle 2 (24 h at 2550 A, 24 h at 1700 A)

from 3360 to 3840 h: load cycle 3 (24 h at 3400 A, 24 h at 1700 A)

Phase 1: maximum temperature of the warmest phase, at the bottom of the middle
      one

Phase 2: maximum temperature of the second phase from the outside at the bottom

Phase 3: maximum temperature at the side of the outside phase

3.8.3.5.5 Result of Comparison Directly Buried

The single-phase GIL was buried beneath the ground at IPH labs (High Voltage Institute in
Berlin), cyclically subjected to high current and high voltage, corresponding to a heating and
cooling phase. The long-term testing of the buried GIL extended to a total of 2500 testing
hours. For the entire duration of test operation, the GIL was subjected during the heating phase
to a current of 4000 A.

Comparisons of the temperatures calculated with the temperatures measured for a period
of 2 weeks with two kinds of cycle occurring and for two locations, for depths of 1 m and
2.9 m, are presented. The calculations agreed well with the test values. The test shows that the
temperature at the circumference of the enclosing tube is not constant and that it is highest at
the top and lowest at the bottom (maximum temperature difference 3°C). Heat is transferred
from the inner conductor to the enclosing tube by natural convection and results in higher
temperatures at the top. This difference does not come out clearly in the calculations since
convection was not taken into account in this calculation [78].

A simulation of an application case of a two-GIL system buried underground shows that
after 6 months of service, the steady state is not fully reached. The middle phase is the warmest
one and the maximum temperature for each phase is at the bottom and at the side [78].

3.8.3.6   Results of Long Duration Test Directly Buried

The long-duration test of the directly buried GIL proved the practical applicability for long-
distance power transmission solutions. The dielectric behaviour of the N_2/SF_6 gas mixture is
reliable and shows long-time stability on the part of the insulation capability. The low thermal
heat-up of the GIL with temperatures below 50°C at 3200 A continuous current (at 400 kV
rated voltage this is equivalent to more than 2000 MVA power transmission capability) has
proven the GIL as a system for transmission of high power ratings with low electrical losses
and high reliability.

The mechanical layout of the GIL is reliable in cases of internal arc faults, with no external
harm or influence to the environment despite all other thermal or electrodynamic forces applied
to the GIL. This was proven with the high ratings of the test sequences.

The GIL passed all the above-mentioned test series successfully and has proven its excellent performance and high reliability, including its secondary equipment. Therefore, from a
technical viewpoint, GIL is an alternative underground transmission system and has many
operational advantages.

The results found by the long duration tests on GIL have proven the expected technical,
economical and environmental aspects under realistic conditions on site. Some of the below
listed key criteria have been found and are worth to mention [38]:

- GIL can transmit 2000 MVA over very long distances.
- Much lower transmission losses than in the case of overhead lines or cables.
- Even over long distances, no reactive power compensation is required in contrast to cables.
- No external influence in case of an internal failure.

The tunnel-laid and directly buried GIL with N_2/SF_6 gas mixture passed all type tests and long-duration testing with excellence. No failure occurred during the total tests, which makes the GIL a very reliable and economical high power rating transmission system.

3.8.4 Long-Duration Test Results

The test results of the long-duration tests showed that GIL with N_2/SF_6 gas mixtures is a reliable technology. GIL proved the ability to bring high power rating of overhead lines directly under ground without reducing the power rating.

The use of N_2/SF_6 gas mixtures is restricted to insulation only; because of the reduced arc-distinguishing feature, no switching is allowed in gas mixture compartments. The gas compartment must be free of switching operations, circuit breakers, disconnectors and earthing switches. The use of N_2/SF_6 gas mixtures and gas handling on-site have been proven to be reliable and practical in handling processes [4].

Development Programme
The development programme has two basic sequences:

1. Type tests on GIL to prove the high-voltage design.
2. Long-duration testing on a 70 m tunnel and a 100 m directly buried GIL test set-up.

The development programme was started in 1995 in cooperation with Electricité de France (EDF) for the first sequence. On the basis of the results of that study, long-duration testing on the tunnel has been carried out in cooperation with the Bewag, utility of Berlin.

Results of the Type Test
The results of the type test can be concluded as follows.

The insulation capabilities of N_2/SF_6 gas mixtures have been under investigation for some decades. Most of the measurements have been carried out on small-scale test set-ups, but could clarify the basic behaviour of N_2/SF_6 gas mixtures [33]. The N_2/SF_6 gas mixture ratio of 80% N_2 and 20% SF_6 was found to be reliable and applicable.

The GIL passed all test requirements given for power frequency, lightning impulse and switching impulse test voltages in accordance with IEC 62271-204 [54]. The high-current performance of 63 kA short-circuit current and high-rated currents (4000 A) for the thermal behaviour have been successfully tested. The electromechanical forces related to the high currents are a basic strength of the coaxial single-phase insulated GIL.

The arc fault tests with currents up to 63 kA have shown the following results: no external influence, because no burn-through of the enclosure or opening of a pressure release device. The main reason is the distributed arc footprint during arc time.

The arc-distinguishing capability, which is necessary for switching operation, is reduced compared to pure SF_6. For GIL the arc behaviour in N_2/SF_6 gas mixtures is an advantage because of no burn-through (for GIS and for circuit breakers it is a disadvantage because of reduced switching capability) [146].

Results of Long-Duration Tests

For the long-duration test in a tunnel, a 70 m-long concrete tunnel including a 20° angle has been erected in cooperation with Bewag, Berlin. Under realistic on-site conditions the test set-up has been assembled. The long-duration test programme covers thermal, mechanical and electrical stresses to simulate a lifetime of 50 years.

After the initial commissioning tests the GIL has been applied to a 5 h heating cycle with a current rating 30% above the rated value, so that the maximum temperature of 50°C was reached after 5 h. During the cooling sequence an AC test voltage was applied at 420 kV, which is double the rated value. To reach the start temperature, forced cooling was applied to the tunnel.

In the middle of the 2500 h total test time a simulated repair of the GIL was carried out under on-site conditions. The repair included the replacement of a 5 m-long section and recommissioning within only 3 days. The second sequence of long-duration tests was finally concluded with a final test and the same test values as the commissioning test. During the completed long-duration test no problem occurred with the GIL. The PD measurements based on the sensitive UHF measuring method showed constant low values, close to zero. No ageing effect could be detected.

The GIL has proven its reliability over the simulated lifetime of 50 years and the realistic assembly on-site showed the robust applicability for industrial projects. A first project has been installed at the 400 kV substation of Kelsterbach in Germany in 2010 and is in operation since the beginning of 2011, see clause 2.2.2.4 and 7.2.13.

3.9

Gas Handling

3.9.1 General

The GIL is a gas-filled high-voltage system. The purpose of the gas is electrical insulation. The gas is not moving inside, it is kept fixed in gas-tight compartments for the lifetime of the GIL. The gases used (SF_6 and N_2) are inert and non-toxic. The filling pressure is relatively low, with a typical value of 0.8 MPa. The metallic enclosure is solidly grounded and delivers, because of the wall thickness of the outer enclosure, high personnel safety. The automated orbital welding process makes sure that the connections of the GIL segments are gas-tight for the lifetime of the GIL.

Gas handling is organized in a closed-loop process where the insulating gas is always kept inside containers. The gas-handling process avoids releasing gas to the atmosphere. The use of SF_6 gas is protocolled in the quantity and quality of the gas at every step of the gas handling. IEC 62271-303 [58] and IEC 62271-4 [51, 60] give detailed gas-handling instructions for safe and economic gas handling.

3.9.2 Gas Mixture Handling

For the gas handling of the N_2/SF_6 gas mixture, devices are available for evacuation, separation, storing and filling of the N_2/SF_6 gas mixtures [7, 209]. In Figure 3.9.1 the devices for gas handling are shown.

From the GIL system with a vacuum pump (1) the gas will be pumped out of the GIL system, then the gas is separated (2) with a gas filter into pure SF_6 and the remainder of the N_2/SF_6 gas mixture with low SF_6 content. The rest of the N_2/SF_6 gas mixture will have an SF_6 content of a few percent (1–5%) so it can be stored (3) under high pressure up to 20 MPa in standard steel bottles. A set of steel bottles is shown in photo (3). This set of steel bottles can be extended to take the gas of a 1 km GIL section for storage. The pure SF_6 will be stored (4) in fluid conditions also under high pressure. To fill or refill the GIL system a gas mixing device (5) is available, including a continuous gas monitoring system for temperature, humidity, SF_6 percentage and gas flow. The gas-mixing device uses pure N_2 (6), SF_6 (4) and the stored gas mixture with a few percent of SF_6. The mixing device will adjust the chosen N_2/SF_6 gas percentage used in the GIL (e.g., 80% N_2).

Gas-Insulated Transmission Lines (GIL), First Edition. Hermann Koch.
© 2012 John Wiley & Sons, Ltd. Published 2012 by John Wiley & Sons, Ltd.

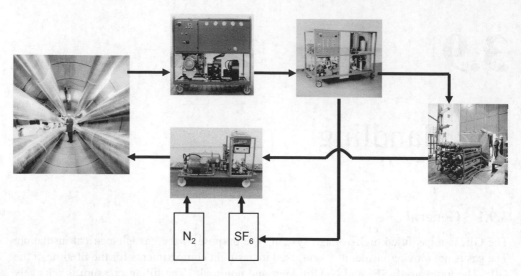

Figure 3.9.1 Gas mixture handling. Reproduced by permission of © DILO

With these gas-handling devices a complete cycle of use and reuse of the gas mixture is available. In normal use the SF_6 and N_2 will not be separated completely because the gas mixture will be reused again. A complete separation into pure SF_6 as used, for example, in GIS can be done at the SF_6 manufacturer's facilities. The recycled gas will fulfil the IEC 60480 [148], IEC 61634 [149] and IEC 62271-4 [51] requirements.

Gas Storage
The concept of N_2/SF_6 gas mixtures is that the gas mixture will stay mixed as long as it is in use. Only the percentages are varied for use in the GIL (e.g., 20% SF_6 and 80% N_2) and for storage (e.g., gas mixture of 1% SF_6 and 99% N_2 and separate pure SF_6). When stored in containers or bottles the nitrogen is compressed to very high pressures of up to 20 MPa to save space for storage. The SF_6 can be stored in fluid conditions at much lower pressure, which reduces the required storage volume. For large gas compartments of length 1 km, several containers are connected together.

Gas Mixture System
To handle the large gas volumes and gas pressures an automated gas-mixing system has been developed, see Figure 3.9.2. This gas-mixing system can produce any mixture ratio with an accuracy $< 1\%$. The right dew point is reached by using AlO_2 filters to dry the gas before filling. The filling temperature is continuously protocolled and stored in the control computer for gas quality control.

The gas-mixing system uses the principle of continuous gas mixture control. This mixing technology allows the gas mixture to be filled in the GIL on the basis of a molecular mixture. This means that the gas mixture is completely mixed and no separation of SF_6 and N_2 will occur.

Figure 3.9.2   Gas-mixing device for on-site application. Reproduced by permission of © DILO

3.9.3   Conclusion

For gas mixtures of SF_6 and N_2, today standard filling and mixing devices are available. The gas-mixing process is automated and the complete filling process is protocolled by a computer system.

The insulating gas of the GIL is always kept inside containers so that the release to the atmosphere is reduced to negligible volumes.

The gas-handling machine cleans the gas mixture when treated with filters, so that the SF_6/N_2 gas mixtures can be reused.

The gas-handling devices also allow the handling of large gas volumes, e.g. GIL gas compartments of 1 km length, in reasonable time. A filling process can be carried out within one day.

If the SF_6/N_2 gas mixture is not needed any longer, it can be sent to the SF_6 gas manufacturer for complete separation and reuse of the SF_6 in the GIS equipment. The quality of reused gas is regulated by IEC standards.

Conclusion

3.10

Commissioning and On-Site Testing

For the commissioning of a GIL, a wide scale of electrical and mechanical testing is necessary. Tests after the assembly is completed prove the correct installation of all elements of the GIL.

The on-site testing is not a repetition of the type tests of the equipment. On-site testing proves the correct assembly and installation of the equipment, not the design. This is done with design or type tests. That is the reason why 80% of the type test voltage is used for testing. The values for a 420 kV GIL are given in Table 3.10.1.

The on-site high-voltage tests are carried out together with PD measurements. The 80% type test voltage needs to be PD-free, which means a value below 5 pC.

In Figure 3.10.1 the power frequency on-site test is shown using resonance frequency test equipment connected by air/SF_6 bushings. The high-voltage test transformer (on the right) is connected by a coupling capacitor (next to the test transformer) to the gas-insulated equipment via air/SF_6 bushings. The coupling capacitor is chosen to reach resonance frequency of the inductance of the test transformer with the capacitance of the coupling capacitor and the capacitance of the gas-insulated equipment. This resonance test method has the advantage of relatively small test equipment by using the resonance voltage increase to reach the test voltage values.

In Figure 3.10.2 a gas-insulated test transformer is shown connected to the gas-insulated system. This test method is used in cases where only limited space is available. This is the case when on-site tests are carried out inside a building or in a tunnel. For GIL, several kilometres of phase length can be on-site tested with such equipment. To allow this, the completed transmission line is separated into some kilometre long (typically 1–2 km) test sections. This test method allows testing the complete transmission line with the high on-site test voltages of 80% of the type test voltages as shown in Table 3.10.2.

In addition to the high-voltage tests, the tests listed in Table 3.10.2 are part of the commissioning.

Gas-Insulated Transmission Lines (GIL), First Edition. Hermann Koch.
© 2012 John Wiley & Sons, Ltd. Published 2012 by John Wiley & Sons, Ltd.

Table 3.10.1   On-site high-voltage test values for a 420 kV GIL

	Type test (kV)	On-site test (kV) (80% level)
Power frequency 1-min test	630	504
Lightning impulse voltage (1.2/50 μs)	1425	1140
PD-free voltage	520	504

Pressure Test

The enclosure pipe of the GIL, the enclosures of the disconnecting and angle units and the gas-tight insulators are designed following the rules of pressure vessel standards. The GIL will be assembled on site to form the transmission line and the standards and regulations require an on-site pressure test to prove the correct assembly and safety of the pressurized equipment.

Depending on the local regulations and the type of quality control of the welded joints, the pressure test will be carried out with 1.1 or 1.3 times the maximum operating pressure before commissioning. This test is usually done with dry air or nitrogen. In some countries the test needs to be carried out with an eye witness representing the responsible authority.

Gas-Tightness Test

Any O-ring seal at the disconnecting, compensator and angle units is checked for gas-tightness using an SF_6 sensor device ("gas sniffer"). The welded joints are checked by a 100% ultrasonic test for each weld.

Figure 3.10.1   Power frequency on-site test using resonance frequency test equipment connected by air/SF_6 bushings. Reproduced by permission of © Alstom Grid

Figure 3.10.2   Gas-insulated test transformer connected to the gas-insulated system

Table 3.10.2   On-site tests for commissioning

Pressure test	The pressure test is defined in EN standards and national regulations and proves the pressurized GIL before operation.
Gas-tightness test	Welding seams and each flange connection is 100% tested for gas-tightness.
Gas mixture	The gas mixture percentage of SF_6 (20%) and N_2 (80%) is checked and protocolled by gas-handling devices.
Gas pressure	The pressure of the gas mixture is checked relative to the temperature to guarantee the required gas density (gas filling density).
Gas humidity	The dew point of $-20°C$ is checked to guarantee low humidity in the gas mixture.
Gas density	The gas density is measured for each gas compartment and a warning signal is given to the control centre by the gas density meter.
Passive corrosion protection test	When directly buried, the passive corrosion protection is tested by 10 kV/1 min. When laid in a tunnel or above ground, usually no corrosion protection is needed.
Active corrosion protection test	For directly buried GIL an active corrosion protection is added and tested before commissioning.
Resistance of the conductor	The resistance of the main circuit is tested according to standards [4.10-1] to prove the correct connection.

Gas Mixture

The gas mixture is controlled by the gas-mixing system. During the filling process the SF_6 portion in percent, the volume per minute, the gas temperature during filling and the humidity are measured and protocolled for each gas compartment.

The gas-filled GIL will be checked for the correct gas mixture (e.g., 20% SF_6 and 80% N_2) during the filling process and after some days to confirm the mixture correctness.

Gas Pressure

The gas pressure is measured and the gas density is calculated using the gas temperature. For this, a precise gas pressure meter is used and each gas compartment is controlled separately.

Gas Humidity

The dew point of the gas humidity needs to be below −20°C to avoid humidity reaching the dew point. Gas dew on the surface of the insulators may cause discharges and needs to be avoided in any case within the operational temperature range.

Gas Density

The gas density is the important value for electrical insulation. Therefore, the gas density of each gas compartment is monitored by a gas density meter. This gas density meter gives monitoring signals to the control centre. Usually two stages are signalled:

    Stage 1: gas loss warning
    Stage 2: automatic switch off

In some cases two stages of gas loss warning are given. The warning level is usually reached after 10 kPa of pressure loss. Automatic switch off is initiated when the gas pressure is below the minimum operating pressure. This may be the case when the gas pressure is 0.1–0.2 MPa below the filling pressure. The values are recommended by the IEC Standard 62271-204 [54].

Passive Corrosion Protection

The passive corrosion protection coating is checked on site for the welding section by a high-voltage test of typically 10 kV for 1 minute. The enclosure pipes are coated in the factory and the high-voltage test of the corrosion protection is carried out there. The welding area for jointing on site is coated by a corrosion protection and also the high-voltage test is carried out there.

Active Corrosion Protection

In addition to the passive corrosion protection an active corrosion protection is used to avoid corrosion. The GIL enclosure pipe is kept at a certain voltage (about −1 V) and a sacrificial electrode will corrode instead.

The on-site test is to measure the GIL enclosure's electrical potential and the current in the protection circuit, which should be zero if no crack or hole has occurred in the passive corrosion protection coating.

Resistance of the Conductor

The resistance of the conductor is measured to prove the correct jointing of each conductor section. Therefore a measuring current of, for example, 100 A is injected as recommended in IEC standard 62271-204 [54].

4

System and Network

4.1 General

In this chapter the different aspects of integrating a GIL into the network and operating the GIL as part of the power transmission system are put into focus. By its physical nature the GIL is best suited for high-power transmission capability. The high current ratings of GIL are produced by the use of aluminium conductor pipes for enclosures of large diameter. This requires – for mechanical stability – a minimum wall thickness of the aluminium pipes, which leads to large cross-sections and high current-carrying capability. The voltage levels are given by the gas insulation using overpressure and the high insulation capability of SF_6 and N_2. Compared with air-insulated systems, the required insulation diameters of the GIL increase only slightly (e.g., from 500 mm diameter for a 500 kV GIL to 650 mm for an 800 kV GIL). In conjunction with current and voltage ratings, high transmission capabilities of 3000 MVA to 6000 MVA are possible.

To integrate such a strong transmission line into the transmission system, some parameters need to be investigated before installing the GIL. Aspects such as physical line parameters, transmission losses, operation, ageing, safety, maintenance, repair and monitoring are part of these investigations.

4.2 Line Constants of GIL

4.2.1 Theoretical Background

The GIL uses the electrical insulation capability of gases and the current-carrying capability of aluminium pipes to build up a transmission system. The solid insulation of the cast resin insulators makes up less than 0.1% of the volume of the GIL. This makes the GIL electrically similar to an overhead line, only enclosed in an outer enclosure pipe. The outer enclosure keeps away environmental impacts like dust or humidity and offers constant electrical conditions. These electrical conditions are represented by constant physical parameters.

The gas insulation dominates the dimensioning of a GIL. This means that the solid insulators are used far below the theoretical insulation capability by the ratio of ε_r, the relative electrical constant of cast resins, which is in the range of 3.5 to 4.5. Therefore, no electrical ageing will occur during the operational lifetime of the GIL.

Gas-Insulated Transmission Lines (GIL), First Edition. Hermann Koch.
© 2012 John Wiley & Sons, Ltd. Published 2012 by John Wiley & Sons, Ltd.

Table 4.1 Typical values of resistance

Rated voltage (kV)	Diameter (mm)	Wall thickness (mm)	Typical AC resistance per phase meter ($\mu\Omega$)
145/170	240	15	18
245/300	310	12	16
362	380	12	13
420/550	500	10	11
800	630	10	10
1200	760	10	8

The line parameters are related to the dimensioning of the enclosure pipe, the gas insulation and the aluminium conductor enclosure pipes.

4.2.2 Resistance

The resistance is related to the diameter and the wall thickness of the pipes and the materials. For the conductor pipe, electrical aluminium with the highest values for conductivity is used.

Electrical aluminium has a purity of 99.5% or higher. For the enclosure pipe, aluminium alloys are used with high mechanical strength to meet the requirements of the pressurized compartment of up to 0.8 MPa, and the mechanical forces coming from the soil load, thermal expansion and bending. Materials of $AlMg_3$ or $AlMg_{4.5}$ are used. The typical values are shown in Table 4.1 [132].

4.2.3 Capacitance

The capacitance of a GIL is given by its dimension and by the gas for insulation. The solid insulators, with a volume of less than 0.1%, can be neglected. The design is cylindrical, with the conductor pipe inside the enclosure pipe. The dielectric constant of the insulating gas is 1. In Table 4.2 typical values are given for different diameters. Compared with overhead lines the capacitance is about 2 times higher and compared with solid insulated cable the capacitance is about 3 to 4 times lower [132].

Table 4.2 Typical capacitances of GIL

Rated voltage (kV)	Diameter (mm)	Wall thickness (mm)	Typical capacitance per phase meter (pF)
145/170	240	15	59
245/300	310	12	52
362	380	12	53
420/550	500	10	54
800	630	10	45
1200	760	10	42

Table 4.3 Typical inductances of GIL

Rated voltage (kV)	Diameter (mm)	Wall thickness (mm)	Typical inductance per phase meter (μH)
145/170	240	15	0.187
245/300	310	12	0.211
362	380	12	0.210
420/550	500	10	0.205
800	630	10	0.247
1200	760	10	0.260

4.2.4 Inductance

The inductance of the GIL is given by the dimensions of the enclosure and conductor pipe and the solid grounded enclosure. The impact of the insulating gas and the insulators can be neglected. Because of the cylindrical design and the solid grounded enclosure, the inductance is relatively small. In Table 4.3 typical inductances are given for different diameters [132].

4.2.5 Impedance

The impedance of a GIL is given by the inductance and the capacitance. The dominating value comes from the capacitance, which is highest when the GIL is operated at no load. With increasing load the inductance is increasing and at rated current the GIL is close to the natural load value, where inductance and capacitance are almost equal and the transmitted power is almost the real power and the impedance is only resistive.

4.2.6 Surge Impedance

The surge impedance is given by the relationship

$$Z' = \omega\sqrt{L/C}$$

The surge impedance is important for the transient behaviour of the GIL when connected to the transmission network. Transient voltages are reflected at connection points of different surge impedances. The reflection is related to the relationship of the connected surge impedances and needs to be evaluated in the insulation coordination study [132].

In Table 4.4 the surge impedances of GIL, overhead line and cables are given.

4.2.7 Natural Power

The natural power of a transmission system is the value that indicates when the inductive and capacitive values are reaching the same value. At this point, only real power is transmitted. This is the optimum for a transmission system. The natural power of the GIL is at power ratings of about 2000–3000 MVA, depending on the dimensioning, and therefore close to the rated power of the GIL.

Table 4.4 Surge impedance of GIL

Rated voltage (kV)	Diameter (mm)	Wall thickness (mm)	Typical surge impedance (Ω)
145/170	240	15	56.0
245/300	310	12	63.4
362	380	12	62.8
420/550	500	10	61.5
800	630	10	73.9
1200	760	10	78.0

The natural power of a 400 kV overhead lines with a conductor bundle of four conductors is in the range of 700–900 MVA, the rated power of a 400 kV overhead line is in the range of 1800–2000 MVA. The transmission system is operated in the inductive vector segment and capacitive compensation is needed for long-distance transmission. Large capacitor banks need to be installed.

Solid insulated cables have a natural power value of 5000 MVA and therefore operate far below this optimal operation point at natural power. Large coils for inductive compensation are needed. These compensation coils have large dimensions and produce thermal losses [132].

In Table 4.5 the natural power of different transmission systems is given.

4.3 Transmission Losses

4.3.1 General

The transmission losses are an important evaluation criterion mainly for high-power transmission systems. Transmission losses are for the lifetime, which adds up to large amounts of energy and money.

The GIL uses aluminium pipes for the conductor and the enclosure. Because of the gas as electrical insulation, the diameter of the GIL is large (e.g., a 420 kV GIL has 500 mm diameter). The large diameter needs, for mechanical stability reasons, a minimum wall thickness (e.g., a 500 mm enclosure will have a minimum of 6 mm wall thickness). This leads to large cross-sections. In the above-mentioned example the enclosure pipe has a cross-section of 18,700 mm^2.

Because of these large cross-sections of enclosure and conductor pipes, the GIL has low transmission losses. A second source of losses is related to resistive losses in the power factor compensation unit. These are needed to compensate the phase angle due to the capacitive load

Table 4.5 Natural power of 400 kV transmission systems

System	Natural power (MVA)	Rated power (MVA)
GIL: 500 mm enclosure pipe diameter	2000–3000	2000–3000
Overhead Line: Conductor bundle of four conductor wires	700–900	1800–2200
Solid cable: XLPE insulated	4000–5000	800–1200

Table 4.6 Typical values of GIL transmission losses

Rated voltage (kV)	Rated current (A)	Power losses per metre (W)
145/170	2500	120
245/300	3000	150
362	3500	170
420/550	4500	230
800	5000	260
1200	5500	240

of the line. Because of the gas insulation, the capacitive load for a GIL is low and therefore the need for compensation.

4.3.2   GIL Losses

The resistivity of the GIL is shown in Table 4.6 for different voltage levels. Based on a typical current rating, the typical power losses can be calculated. In principle, the higher the voltage ratings the larger the diameters, the lower the specific transmission losses. The absolute values of losses increase with the voltage level because of higher rated currents. In Table 4.6 the values are shown for different voltage levels.

4.3.3   Comparison with Other Transmission Systems

The comparison with other transmission systems is a complex task and evaluates only one aspect of the different features of transmission systems. Here a comparison is made for 400 kV transmission systems as they are used today. In Table 4.7 the technical data for the comparison is given. The investigated transmission systems are overhead line, oil cable, XLPE cable and GIL.

The type of overhead line is standard in Germany using a bundle of four wires of 240 mm^2 of aluminium and 40 mm^2 of steel. The thermal transmission power limit is 1790 MVA, which relates to the current of 2580 A. The oil cable has a copper cross-section of 1200 mm^2 and a thermal power limit of 1120 MVA. Usually such cables are only used as cross-bonded installations because of the limitation of current flow in the cable shield. The same holds for XLPE cables, where the copper cross-section is chosen to be 1600 mm^2.

Two GIL versions are chosen, one for 2000 MVA rated power and the other for 3000 MVA rated power, with a thermal power limit of 2180 MVA and 3190 MVA. The thermal current limits are 3150 A and 4600 A.

In Figure 4.1 the values of transmission losses are shown in a graphic to give an overview of the different transmission systems. The highest losses are with grounded cables. For this reason grounded cables are not used when high-power transmission is required, only cross-bonded cables are used.

The overhead line has higher losses compared with cables and GIL. The cables chosen here are standard 1600 mm^2 cables. Today, 2500 mm^2 cables are also in use and then the losses of cables and GIL are almost the same. The lower cross-section of the cable is compensated by the lower resistance of the copper conductor compared with the aluminium conductor and the layer cross-section of the GIL.

Table 4.7 Technical data for different transmission systems

Type	Overhead line 4 × 240/40Al/St	Oil cable 1200 mm^2 Tunnel-laid		XLPE cable 1600 mm^2 Tunnel-laid		GIL Tunnel-laid	
		Grounded	Cross-bonded	Grounded	Cross-bonded	Ø 520 mm grounded	Ø 600 mm grounded
Thermal power limit $S+n$ (MVA)	1790	–	1120	–	1150	2180	3190
Thermal current limit I_{th} (A)	2580	–	1610	–	1660	3150	4600
Resistance R' (mΩ/km)	30.4	40.8	23.0	35.7	19.0	9.4	6.9
Losses at limit current (W/m)	600 at 2500 A	–	150 at 1650 A	–	190 at 1660 A	380 at 3150 A	440 at 4600 A
Losses at 1600 A (W/m)	230	310	145	275	177	72	53

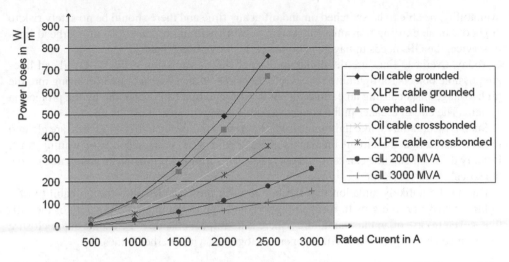

Figure 4.1 Transmission losses of different systems. Reproduced by permission of © Siemens AG

4.3.4 Cooling or Ventilation

For the GIL laid in a tunnel or directly in the ground, the design is chosen so that no additional cooling system is required at the rated current. In a tunnel, the natural convection of the air in the tunnel will transport the heat to the tunnel walls and the tunnel walls have a conductive heat transfer in the surrounding soil to transport the heat coming from the transmission losses. When the GIL is directly buried, the heat coming from the transmission losses is directly conducted to the surrounding soil. The thermal conductivity of the surrounding soil is chosen in such a way as to be below the maximum temperature of the GIL enclosure – which in most cases is 40°C [90, 142].

To increase power transmission, cooling or ventilation is an option. A wide variety of options are possible.

Tunnel-Laid   The first choice is ventilating the tunnel and using ventilation shafts for air exchange in the tunnel. In a second step, the air may be cooled down to increase the heat transportation out of the tunnel.

Buried GIL   The cooling of buried GIL starts with improved soil bedding materials with higher thermal conductivity. In a second step the bedding may be equipped with water pipes to transport the heat to cooling stations along the transmission line.

In any case, cooling and ventilation is costly for the investment and also during the lifetime for maintenance works. Cooling and ventilation should be avoided.

4.4 Operational Aspects

4.4.1 General

The operational aspects from a system operator's view are linked to availability, line switching and safety. From the operational point of view the transmission system should have a high availability, be able to be switched on and off at any time and there should be no safety risk to anybody. In an existing transmission system network with more than 90% of overhead lines in service, the GIL needs to have a good match with overhead lines.

Today, overhead lines are the dominant part of the transmission network. Overhead lines are possible objects for lightning strikes into the wires or the towers, or trees growing into the high-voltage lines causing an earth fault. Both will lead to a trip of the line by the protection system. The circuit breaker in the connected substation will switch off the energy.

Such typical failures, like lightning strikes or ground faults due to tree branches, are cleared in seconds so that the circuit breaker is closed again after some seconds of waiting time. Usually the overhead line can continue operation. This tripping is done in an automated mode, the so-called autoreclosure.

Due to the unlikely situation that the failure is in the GIL (because the probability of a lightning strike or tree growth is much higher), and due to the integrity of safety of the GIL that – even in case of an internal failure – there is no impact outside the GIL, the autoreclosure function can be allowed when GIL is operated together with overhead lines.

4.4.2   Availability

Availability of the transmission line means keeping it in operation at almost any time. Mainly in high-power transmission, the interruption has significant consequences for the power supply. In some cases the loss of a high-power transmission line may end in a regional blackout as we have seen in California, the North East USA or Italy over recent years.

The reason for interruption may come from a failure in the GIL or an interruption in the connected overhead line system. The probability of a failure in the GIL is very low. More than 30 years of experience world-wide, with more than 300 km in operation today, have shown no failure to be reported that leads to a power supply interruption. Failures only occurred during the commissioning process. Experience shows that, once the GIL is in operation, it will operate without failure. No ageing effects have been detected.

The most likely interruptions in operation of the transmission network come from the overhead line and these interruptions are usually short-time interruptions. The reason may be a lightning strike in the line or a ground fault caused by a tree. These interruptions are solved in the network by so-called autoreclosure functions. This means that the line will be shut off for some seconds and then switched on again. In most cases the failure has cleared and the line is back in operation.

The GIL can be operated in the same way as an overhead line using the autoreclosure function. To prove that there is no danger for persons or equipment at the transmission line, the GIL has been tested in the laboratory even to withstand internal failures in the GIL.

The autoreclosure function will automatically close the circuit breaker to energize the line. If the failure is at the overhead line, no risk is seen as the clearance does not impact the surroundings.

To prove the safety of the GIL, different internal arc currents have been tested in the laboratory. In the network, the short-circuit rating is different at locations close to power plants with high currents and more distant locations with lower currents (as shown in Table 4.8).

The arc fault test parameters were set to 63 kA, 50 kA and 10 kA with switching off times of 0.5 s, 0.33 s and 0.1 s (as shown in Table 4.9).

The results of the internal arc tests showed no burn-through or opening of the enclosure. Therefore, no danger for persons or equipment nearby. The pressure rise due to the arc did not

Table 4.8 Short-circuit currents at different locations in the network

63 kA	Typical highest fault current rating in the 400 kV network close to substation and power plants.
50 kA	Typical fault current rating in the 400 kV network at substations away from power plants.
10 kA	Typical fault current rating in the 400 kV network on single transmission lines away from a substation or power plant.

exceed 50%, which is in the range of the pressure design of the enclosure (margin of 250%). After the clearance of the arc according to the given times, the rated voltage was applied to the GIL – which withstood it. This allows continuation of operation, even if a failure happened in the GIL (which is very unlikely).

The application of the autoreclosure function to a combination of overhead lines with GIL is allowed. This increases the availability of the transmission network.

Line Switching The ability to switch the line at any time is important to the operator of the transmission network. The GIL as a gas-insulated system has similar switching behaviour as overhead lines. The overhead lines are usually much longer than the GIL part, therefore, the absolute capacitance of the GIL is small compared with overhead lines. The physical capacitance of the GIL is about double that of an overhead line. Therefore, inrush currents of transmission networks with GIL are similar to those of pure overhead lines. Together with the safety aspect explained before, the GIL can be operated like an overhead line.

Operational Safety The safety aspect in terms of availability is very much linked to the internal arc failure situation. For that safety reason, type tests with internal arc faults have been investigated and delivered good results for safety of persons around the GIL.

A test set-up with major components of the GIL was chosen to verify the arc withstand capabilities of the enclosure in the case of an internal arc. According to IEC requirements, a typical arrangement was chosen. The length of the arrangement was about 25 m, so that the gas pressure rise (because of the heat of the arc) did stay within acceptable limits. GIL installations have large gas compartments, so the gas pressure rise during an internal arc fault will be small (less than 1 bar). The set-up was assembled in the laboratory, and for safety purposes rupture discs have been provided that should not open during the test. Figure 4.2 shows the arrangement at the IPH test labs in Berlin.

In a conclusion of the test made as explained before it can be noted that there was no external impact of the surrounding during and after the internal arc test. No burn-through of the enclosure has occurred and only some micrometers of aluminium material has been melted by the arc on the inside of the enclosure pipe and the outside of the conductor pipe.

Table 4.9 Arc fault test parameters

Arc current	63 kA	50 kA	10 kA
Duration of arc current	0.5 s	0.33 s	0.1 s
Arc voltage	16 kV	16 kV	16 kV

Figure 4.2   Test set-up IPH, Berlin. Reproduced by permission of © Siemens AG

The gas pressure rise inside the enclosure pipe during and after the arcing test was significantly low. Even the rupture disk which has been set about 10% above the filling pressure of
0.8 MPa did not open.

The burning characteristic of the arc in the N_2/SF_6 gas mixture show a smooth behavior
concerning the moving speed and the maximum temperature at the foot point of the arc. With
a wider arc diameter compared to SF_6 the maximum temperature at the footprint is low and
the electromagnetic forces driving the arc in SF_6 are also low in N_2/SF_6 gas mixtures.

The epoxy resin insulator have not been seriously damaged by the arc.

This makes the GIL with N_2/SF_6 gas mixture a safe transmission system which could also
be installed inside of traffic tunnels or on bridges.

All these results speak in favour of safe operation of the GIL; even in the very unlikely case
of an internal arc, the environment is not affected. The results of the arc fault test also showed
that in the case of a tunnel-laid GIL, no personal danger for people in the tunnel could be
expected.

4.5   Ageing

Insulating gases are not ageing. This effect can be seen on overhead lines with a non-ageing
capability to insulate the electrical high voltage against ground potential if the design is chosen
correctly and the electric field strengths are in the given limits. The physical background is
that in gases, the molecules are free to move and are powered by thermal convection. With
this movement, free electrons or ions which are captured by the gas molecules are transported

to the ground potential and released. The movement of the gas molecules inside the GIL gas compartments will balance and distribute electrical charges so that no accumulation can occur which might be dangerous to the insulation system. This is a continuous process and practically no ageing effect can be measured.

Ageing of solid materials depends on the exposed values and limits given by the material. Two ageing characteristics are known for materials:

- thermal ageing
- electrical ageing

Thermal ageing changes the material on a molecular basis. Insulating materials are limited by their classification temperature, which is known for each material. Typical values of this temperature limit for resins are 105°C to 120°C. For the technical design of the GIL, these temperatures give the limits.

Electrical ageing of solid materials starts when electric field strengths are above values of 30 to 50 kV/mm, depending on the type of material. The dielectric design of the GIL is dimensioned not by the solid insulators but by the gas insulation, which is a factor of 3–4 lower than the values of the insulators. This allows a maximum of about 10 kV/mm for the GIL design and, therefore, this value is far below the limits for electrical ageing.

A series of long-duration tests on gas-insulated systems have been carried out in laboratories around the world to prove the reliability of casted epoxy resin insulators. The measurements from all of the different technical designs of different manufacturers point to the conclusion that casted epoxy resin insulators are practically not ageing. The results are published in a technical report [195].

4.6 Internal Arc Fault

4.6.1 General

The internal arc fault is a very seldom event in the operational life time of the GIL, but it can occur and therefore is part of the design and the type tests of GIL. In more than 40 years such an internal arc did not happen at GIL in operation, as far as reports are public. The Internal arc test for GIL has the requirement not to impact the surrounding and keep all effect related to the arc (hot gases and melted aluminium) inside the enclosure pipe.

To reach this one principle requirement is the large size of gas compartments. Minimum length of some 10 m are required. In praxis the gas compartments are much bigger as several 100 m up a kilometer.

4.6.2 Passive Protection

The passive protection from the possible impact of the internal arc fault is given by the solid metallic enclosure. The internal arc fault generates hot gases and increased internal pressure. The internal arc test proves that the GIL enclosure pipe can withstand these forces, see Section 3.2.

An additional advantage beside the solid metallic enclosure is given by the physical behaviour of N_2/SF_6 gas mixture when the footprint of the arc tends to be much larger than in pure SF_6, and therefore the temperatures at the footprint are much lower. A burn-through of the

Figure 4.3   Impact of an internal arc 63 kA/500 ms. Reproduced by permission of © Siemens AG

enclosure can be avoided and external impact of the internal arc is given. In the arc fault tests the result of a 63 kA arc current for 500 ms is shown in Figure 4.3. Only a few micrometres of aluminium material have been melted. The enclosure pipe can withstand the internal pressure and no gases are released to the atmosphere.

The gas compartments are limited by gas-tight insulators to limit also the internal impact of the arc. This will reduce the repair length which needs to be replaced after such an event. As said before, the design is made such that an arc fault does not occur, as has been proven over 40 years of operation [172].

4.6.3   Arc Location

The arc location system monitors the GIL continuously, and in case of an internal arc it gives the precise position of the occurrence. For this, the arc location system calculates the travelling time of the electric travelling wave which is initiated by the arc. With the length of the GIL and the travelling speed of the internal travel wave known, the location can be calculated from the time difference between the travelling waves needed to reach each end of the GIL. The accuracy of this calculation is 25 m.

This arc location is important for a quick find and repair of the section in the GIL, see Section 3.3.13.3 for a technical explanation of the arc location system.

4.7   Maintenance

From the network operator's point of view, the maintenance of the GIL has an impact on its availability. There is no real maintenance required for a GIL because it is a passive system, with the only purpose of power transmission. No switching or breaking functions are involved in the GIL. Nevertheless, during operation time some activities are recommended:

- visual inspection of the tunnel or shafts,
- checking of the corrosion protection system,
- checking of the gas density meters.

These visual checks are recommended at long time intervals, e.g. once a year. The monitoring of the right gas pressure and density is done by electrical signals for each gas compartment, and is available at the control centre [111, 165].

4.8 Repair

Repair of a GIL in operation needs time, and the transmission line is out of service. To prove the repair process of the GIL during long-duration tests, a repair has been carried out for the tunnel-laid and directly buried GIL (see Section 3.8 for the long-duration test where repair is explained). The repair time for a tunnel-laid GIL may be three days, and for the buried GIL less than about a week. The repair process can be made shorter when repair tools, gas storage containers, prepared replacement GIL sections and an activity plan for all work is available before the repair is needed [60].

4.9 Personnel Safety

The metallic enclosure makes the GIL a very safe transmission system, where even in the unusual case of an internal arc no external influence occurs on the surrounding. This safety feature makes the GIL the only underground transmission system which can be added to a public traffic tunnel like a railway, subway or street tunnel [151].

Touch Voltage   A second safety aspect concerns touch voltages. To avoid dangerous touch voltages, the GIL is solidly grounded at each possible earthing possibility.

The tunnel-laid GIL is grounded through the steel structures, the reinforcement steel in the tunnel concrete walls and the connected earthing system connected to the tunnel. This means that each person standing in the tunnel and touching the metallic enclosure of the GIL has the same electrical potential.

In case of directly buried GIL, the metallic enclosure is coated for corrosion protection and can therefore not be touched. The metallic enclosures in the shafts are connected to the shaft earthing system. In the same way as in the tunnel, each person in the shaft has the same potential as the metallic enclosure.

Fire Load   The GIL is made of aluminium pipes, cast resin insulators and non-flammable insulating gas. These materials do not add any fire risk to the tunnel installation or when buried underground. The internal arc will be kept inside and does not add a fire load to the surrounding.

4.10   Insulation Coordination

4.10.1   General

The insulation coordination of the GIL design, and the transmission network requirements, are the electrical basis for the integrity of the whole system. The network requirements come from the electrical stresses which occur in the network during normal operation (e.g., switching) or in case of abnormal stresses (e.g., lightning strike into a connected overhead line or an earth fault current in the network). The high reliability of the GIL over the last 30 years of operation in the network in many applications [152] gives practical proof of its reliability as a transmission network. The new development of GIL using gas mixtures follows the same path of high reliability which has been proven in long-term tests [138] for directly buried GIL and tunnel-laid GIL [47]. This positive experience was the input for the development of a directly buried GIL and its long-term testing in a prototype [152]. The first application of this second generation of GIL laid in a tunnel and using an N_2/SF_6 gas mixture has been installed in a project at the Geneva airport [44–46].

In the following, the overvoltage stresses of typical GIL applications in the transmission network, insulation coordination, required test voltage and verification in a test set-up are studied.

4.10.2   Overvoltage Stresses on Typical GIL Applications

4.10.2.1   Basic Arrangement

As an example, for typical GIL applications the basic arrangement shown in Figure 4.4 has been used. It represents the connection of 400 kV overhead lines of different design by a GIL of 1 km (respectively 10 km) length. The overhead line on the left-hand side is protected by two shielding wires along its full length. The height of the last three towers (L1 to L3) is about 65 m, their maximum tower footing resistance about 7.5Ω (L2). The overhead line on the right-hand side has two shielding wires between towers R1 and R3 and only one on its further length. The height of these towers is about 45 m and their footing resistance about 5Ω [91].

4.10.2.2   Maximum Stresses by Lightning Strikes

Based on these configurations of the overhead lines and the lengths of their insulator strings, the following maximum stresses by lightning strikes have been evaluated and used for the calculation of maximum overvoltage stresses on the GIL [38].

Left-hand side overhead line

– remote strike:	2000 kV
– nearby direct strikes:	35 kA at locations N1 and N2 (near towers L3 resp. L1)
– strike to towers:	200 kA

Right-hand side overhead line

– remote strike:	2100 kV
– nearby direct strikes:	18 kA at location N3 (near tower R1) and 50 kA at location N4 (near tower R3)
– strike to towers:	200 kA

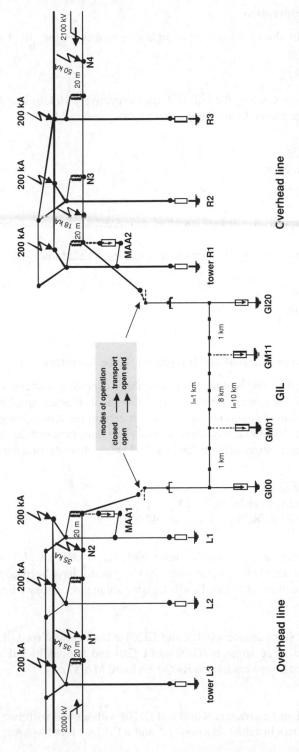

Figure 4.4 Basic arrangement, overhead line/GIL/overhead line. Length of GIL: 1 km and 10 km. Reproduced by permission of © Siemens AG

4.10.2.3   Modes of Operation

The basic arrangement allows calculation of lightning strikes on the GIL for the following modes of operation.

Transport
Both overhead lines connected by the GIL of 1 km (respectively 10 km) length.
  Overvoltage stresses caused by lightning strikes to the overhead line:

- on the left-hand side
- on the right-hand side

Open end
The GIL of 1 km (respectively 10 km) length is connected at one end to the open bay of a substation.
  Overvoltage stresses caused by lightning strikes to the overhead line:

- on the left-hand side
- on the right-hand side

4.10.2.4   Application of External and Integrated Surge Arresters

To protect the GIL against high lightning and switching overvoltage stresses, external surge arresters located at the last towers of the overhead lines as well as encapsulated metal oxide surge arresters immediately connected to the GIL at certain locations (integrated surge arresters) can be applied. For 400 kV systems with an earth fault factor of less than or equal to 1.4, the special integrated surge arresters have the following characteristic data:

- rated voltage $U_r = 322$ kV
- continuous operating voltage $U_c = 255$ kV
- residual voltage at 10 kA (8/20 µs) $U_{10kA} = 740$ kV

As external metal oxide surge arresters, those with $U_r = 360$ kV, $U_c = 288$ kV and $U_{10kA} = 864$ kV, as commonly used in German 400 kV systems, have been taken into account.
  The following applications of external and integrated surge arresters have been investigated.

GIL of 1 km length
Two sets of integrated surge arresters (GI00 and GI20) at both ends of the GIL.
Two sets of integrated surge arresters (GI00 and GI20) and two additional sets of external surge arresters installed on towers L1 and R1 (MAA1 and MAA2).

GIL of 10 km length
Two sets of integrated surge arresters (GI00 and GI20) without and with two additional sets of external surge arresters installed on towers L1 and R1 (MAA1 and MAA2).

Table 4.10 Maximum overvoltage stresses
GIL of 1 km length, mode of operation: transport
Integrated surge arresters GI00 and GI20 and external surge arresters on towers L1 and R1
(MAA1 and MAA2)

Lightning strike		Maximum overvoltages (kV) distance from GIL end on left-hand side			
Location	Peak value	0–250 m	250–500 m	500–750 m	750–1000 m
remote L nearby direct	2000 kV	722	708	722	722
N1	35 kA	528	505	545	673
N2	35 kA	938	952	942	842
L3	200 kA	no backward flashover			
tower L2	200 kA	690	702	704	702
L1	200 kA	no backward flashover			
remote R nearby direct	2100 kV	797	798	787	777
N3	18 kA	872	889	890	823
N4	50 kA	727	705	722	636
R1	200 kA	no backward flashover			
tower R2	200 kA	no backward flashover			
R3	200 kA	no backward flashover			

→ 904 kV

Four sets of integrated surge arresters (GI00, GI20, GM01 and GM11) at both ends of the GIL and at a distance of 1 km from both ends, without and with two additional sets of external surge arresters installed on towers L1 and R1 (MAA1 and MAA2).

4.10.2.5 Results of Calculations

For each mode of operation, both lengths of GIL and the different possibilities of surge arrester application, the maximum overvoltage stresses depending on the distance from the left-hand side end of the GIL have been calculated for all kinds of possible lightning strikes. Tables 4.10 and 4.11 show examples of the results for "transport" operation of GIL of 1 km (respectively 10 km) length. Figures 4.5 and 4.6 illustrate the shape of overvoltage stresses along the GIL for direct strikes of 35 kA to location N2 for both lengths.

In all cases the maximum stresses are caused by nearby direct strikes to line conductors at location N_2 (line on left-hand side) and at location N3 (line on right-hand side). For the various possibilities of surge arrester application, the maximum lightning overvoltage stresses are compiled in Table 4.12 for the GIL of 1 km length and in Table 4.13 for the GIL of 10 km length.

4.10.3 Insulation Coordination for GIL

At least up to a length of some tens of kilometres lightning overvoltage stresses are decisive for the insulation coordination of a GIL, since stresses by switching overvoltages at these lengths will be much lower than on overhead lines – because of the low surge impedance of

Table 4.11 Maximum overvoltage stresses
GIL of 10 km length, mode of operation: transport
Integrated surge arresters GI00 and GI20 and external surge arresters on towers L1 and R1
(MAA1 and MAA2)

Lightning strike		Maximum overvoltages (kV) distance from GIL end on left-hand side			
Location	Peak value	0–500 m	500–5000 m	5000–9500 m	9500–10,000 m
remote L	2000 kV	533	540	707	716
nearby direct					
N1	35 kA	514	464	485	638
N2	35 kA	915	965	983	935
L3	200 kA	no backward flashover			
tower L2	200 kA	492	487	695	696
L1	200 kA	no backward flashover			
remote R	2100 kV	539	556	727	731
nearby direct					
N3	18 kA	885	890	826	761
N4	50 kA	696	611	518	563
R1	200 kA	no backward flashover			
tower R2	200 kA	no backward flashover			
R3	200 kA	no backward flashover			

→ 904 kV

about 60Ω compared with that of an overhead line (about 300Ω). The insulation coordination of a GIL of up to 10 km length considered here is therefore based on the maximum stresses by lightning overvoltages.

According to the substantial experience already gained with design tests and on-site tests of huge gas-insulated substations, the following procedure for the selection of test voltages for on-site tests on GIL sections of up to 1 km length and of type tests on a representative length of GIL are proposed.

On-site tests
These tests have to verify that after laying and assembling the GIL is free from irregularities. Taking into account the safety factor of $K_s = 1.15$ according to IEC 60071-2 [222], a withstand voltage of

$$U_w \geq 1.15\, U_{LEmax}$$

(with U_{LEmax} = maximum overvoltage stress from the calculations) should be verified by these tests.

Type tests (design tests)
The on-site test voltage, on the other hand, corresponds to 80% of the required rated lightning withstand voltage U_{rw} when type testing a representative length of GIL. At these tests single flashovers on self-restoring insulation are permitted according to IEC regulations.

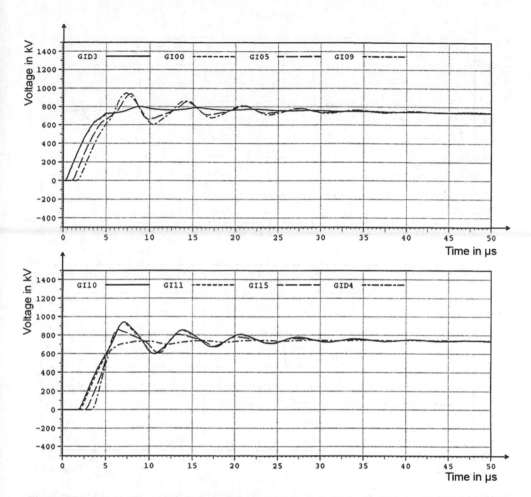

Figure 4.5   Shape of overvoltage stress of 1 km GIL. Reproduced by permission of © Siemens AG

4.10.4   Required Test Voltages

The required test voltages are explained, depending on surge arrester application and the mode of operation.

The above-mentioned procedure results in the following requirements for on-site test voltages and maximum permitted overvoltage stresses on the 400 kV GIL:

Design test U_{rw}	On-site test U_w	Maximum overvoltage stress U_{LEmax}
1425 kV	1140 kV	$\leq$ 990 kV
1300 kV	1040 kV	$\leq$ 904 kV

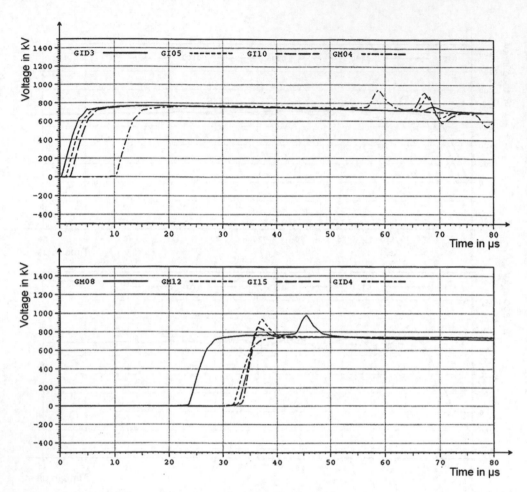

Figure 4.6 Shape of overvoltage stress of 10 km GIL. Reproduced by permission of © Siemens AG

According to the results of the overvoltage calculations and the measures provided for over-voltage protection, the following rated lightning impulse withstand voltages for type tests of the 400 kV GIL considered here have to be applied:

- GIL of 1 km length: Mode of operation: transport and open end (see Table 4.12)
  Overvoltage protection applied: 2 sets of integrated surge arresters (GI00 and GI20) and 2 sets of external surge arresters installed on towers L1 and R1 (MAA1 and MAA2)
  Required lightning impulse withstand voltage: 1425 kV
- GIL of 10 km length: Mode of operation: transport (see Table 4.13)
  Overvoltage protection applied: 2 sets of integrated surge arresters (GI00 and GI20) and 2 sets of external surge arresters installed on towers L1 and R1 (MAA1 and MAA2)
  Required lightning impulse withstand voltage: 1425 kV

Table 4.12 Maximum overvoltage stresses depending on mode of operation and number of surge arrester sets – GIL of 1 km length

	Number of surge arrester sets at:		Maximum overvoltages (kV) caused by nearby direct strokes to:	
Mode of operation	GIL	Tower L1/R1	N2 (35 kA)	N3 (18 kA)
Transport	2	–	1013	913
	2	2	952	890
Open end	2	–	1066	958
	2	2	989	938

— > 990 kV
--- > 904 kV

4 sets of integrated surge arresters (GI00, GI20, GM01, GM11)
Required lightning impulse withstand voltage: 1300 kV
• GIL of 10 km length: Mode of operation: open end (see Table 4.13)
Overvoltage protection applied: 4 sets of integrated surge arresters (GI00, GI20, GM01, GM11)
Required lightning impulse withstand voltage: 1300 kV [91]

The calculations have shown that GIL can be well protected against high overvoltage stresses and that their lightning impulse withstand voltage can be reduced, when additional integrated surge arresters are applied.

Table 4.13 Maximum overvoltage stresses depending on mode of operation and number of surge arrester sets – GIL of 10 km length

	Number of surge arrester sets at:		Maximum overvoltages (kV) caused by nearby direct strokes to:	
Mode of operation	GIL	Tower L1/R1	N2 (35 kA)	N3 (18 kA)
Transport	2	–	996	904
	2	2	983	890
	4	–	867	837
	4	2	842	829
Open end	2	–	1048	950
	2	2	1035	940
	4	–	902	883
	4	2	893	877

— > 990 kV
--- > 904 kV

Figure 4.7　420 kV GIL prototype at the Siemens test laboratory. Reproduced by permission of © Siemens AG

4.10.5　Verification of the Calculated Data

Results of the Type Testing　Figure 4.7 shows a test set-up to verify the dielectric, dynamic and thermal impulse current and the arc fault withstandability.

The results of the type test can be concluded as follows.

The insulation capability of N_2/SF_6 gas mixtures has been under investigation for some decades. Most of the measurements have been carried out on small-scale test set-ups, but clarify the basic behaviour of N_2/SF_6 gas mixtures [59]. For industrial use the SF_6 content could be below 20%, because more than 70% of the absolute insulation capability is reached. The minimum SF_6 content should be above 10%, because of the steep increase of the insulation capability with gas mixtures between zero and 10% of SF_6.

The test set-up also passed all test requirements for lightning and switching impulse voltages in accordance with the IEC requirements.

The high-current performance (63 kA) for thermal behaviour has been successfully tested, and is a basic strength of the coaxial GIL system if single-phase insulated.

The arc fault tests with currents up to 63 kA have shown the following results: no external influence, because no burn-through of the enclosure or opening of a pressure release device occurs. The main reason is the distributed arc footprint during arc time.

The arc-distinguishing capability, which is necessary for switching operation, is reduced compared with pure SF_6. For GIL the arc behaviour in N_2/SF_6 gas mixtures is an advantage because of no burn-through, for GIS and circuit breakers it is a disadvantage because of reduced switching capability [59].

The test requirements for the long-duration test are shown in Table 4.14.

Table 4.14   Test requirements of the long-duration test

• Type test values for commissioning, recommissioning and final tests	
– Short-time power frequency voltage with PD measurement	630 kV/1 min
– Switching impulse voltage	1050 kV
– Lightning impulse voltage	1300 kV
• Long-duration test values	
– 7 h power frequency voltage	480 kV
– 5 h heating current	3200 A
– After each 480 h sequence: switching impulse voltage	1000 kV

As the GIL is an electrically closed system, no lightning impulse voltage can strike the GIL directly. Therefore, it is possible to reduce the lightning impulse voltage level by using surge arresters at the end of the GIL. The so-called integrated surge arrester concept allows reduction of high-frequency overvoltage by connecting the surge arresters to the GIL in the gas compartment [37, 61].

4.11   System Control

4.11.1   Introduction

To operate the GIL in a network, some system control is necessary: gas density, partial discharge and temperature are measured and monitored. The gas density needs to be monitored during operation, partial discharge measurements are only needed for the commissioning process and temperature measurements are only needed at specific locations along the line where thermal overheating may be possible [60].

4.11.2   Gas Density Monitoring

To measure the gas density, a temperature-compensated pressure gauge is used which gives a warning signal when a defined gas density level is reached. The signal is connected to the control centre for each single gas compartment.

The dielectric property is related to the density inside the enclosure pipe. The gas density depends on temperature and pressure. For practical reasons the temperature is measured and compensated using the same ambient temperature of the measured gas volume. The so-called "temperature-compensated pressure gauge".

The measuring result is transmitted to the control centre and used in the automated control scheme. The pressure values are not indicated, instead only contact signals are used to give a "warning of gas loss" and "switch off signal". The warning of gas loss is given when the density is below a given value, usually 5–10% below the minimum filling pressure. In GIL with gas compartments these values are sometimes set to 3–5% below the minimum filling pressure.

The second stage of gas density signal is set for switching off the line when the gas density is below the value of safe dielectric insulation. This value depends on the design and may be in the range of 30–50% of the minimum gas density. When such a low value is reached, the gas density monitoring system gives a signal to the control system to shut off the GIL to avoid any internal flashover.

The gas density monitoring system is a continuous measuring system of the GIL to ensure safe operation and part of the system control.

4.11.3   Partial Discharge Measurement

The partial discharge measurement is used during the commissioning process when the on-site test voltage is applied. This very sensitive measurement system is connected by internal antennas to detect partial discharge and to give integrity of the electrical insulation system.

Once these commissioning tests have been successfully completed, the partial discharge measuring system can be disconnected [207, 208, 213, 214, 215, 219].

Experience with gas-insulated technology (GIS and GIL) over the last 40 years has shown that once a gas compartment is free of partial discharges it will stay that way. Only very few failures are related to increasing partial discharge intensities in gas-insulated systems. Once the partial discharges disappear during the high-voltage commissioning process, there is no need for continuous partial discharge measurement of the GIL.

The UHF internal antenna to measure the electromagnetic radiation inside the GIL will stay in the GIL, only the external measuring equipment needs to be connected when required.

4.11.4   Temperature Measurement

During operation the GIL may have different thermal locations because of external circumstances. Road underpasses, deeper laying requirements, temporary soil dryness or other reasons may cause thermal locations with higher temperature. Control of such locations – which are identified when the GIL route is studied – can be monitored by external temperature sensors to avoid overheating of the GIL [78, 90].

The design and layout of a GIL is made according to the worst case scenario of operation. The maximum temperatures are calculated for the laying conditions in a tunnel, trench, above ground or directly buried for operational conditions. Maximal load, maximal ambient and soil temperature and expected load cycles are the calculation criteria for the thermal dimensioning of the GIL. If the conditions are not changing, no temperature measurements on a continuous basis are required. Only if conditions change is it necessary to take control of the outside and inside temperatures of the GIL. Higher loads than expected, or temporary overload conditions, may be a reason for condition change.

4.11.5   Overview of GIL Monitoring

The different monitoring devices and functions of gas density, partial discharge and temperature sensors – together with the arc location system – are linked together in one GIL monitoring system which is usually located in the control centre [111, 165].

In Figure 4.8 the GIL monitoring system of the control centre is shown. For two three-phase GIL systems (green colour) the six single pipes have their own sensors for monitoring. The gas density gauges and temperature sensors (blue colour) collect the measurement data and link it

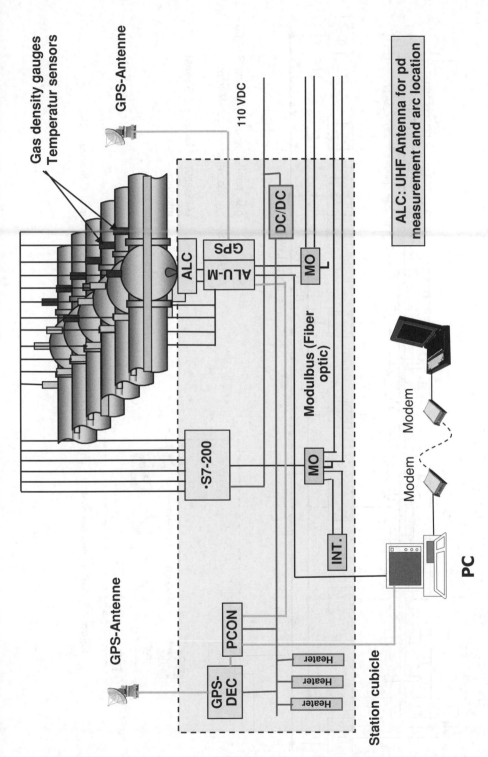

Figure 4.8 GIL monitoring system of the control centre. Reproduced by permission of © Siemens AG

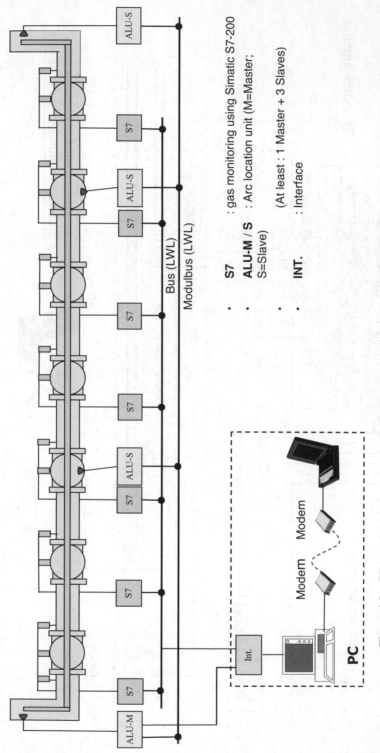

Figure 4.9 GIL monitoring system – physical system overview. Reproduced by permission of © Siemens AG

via a data communication bus "Modulbus (LWL)" to the control centre. The partial discharge measurements and the arc location sensoring are made by the arc location unit (ALC). This information is pre-processed at the GIL location and then transmitted to the control centre.

The gas density gauges and temperature sensors measure the values for each gas compartment. The gas density gauges give the present signals for "alarm of minimum gas density" and "shut off because of too low gas density" to the collecting computer S7-200. These signals are part of the control system of the GIL. The gas temperature is measured and transmitted as an analogue measuring value to the collecting computer S7-200. The temperature values are for information only and do not have automatic influence on the control system. Via a fibre-optic module bus the data is connected to the operation centre.

The ALC is connected at the end of the transmission line or in sections at 3–5 km distance to give precise information about the internal location of an arc failure. Connected to two GPS antennas at the end of the line or section, the measured ALC signal identifies the location of the arc with an accuracy of ± 25 m by calculating the surge travelling time of the electromagnetic impulse inside the GIL. This data is also transferred to the control centre and the location is indicated at the control monitor. All sensors are connected to disconnecting units.

In Figure 4.9 a physical principle is shown for the positions of the monitoring sensors. The GIL is divided into gas compartments of about 1 km length and the gas compartments are separated by disconnecting units. Each compartment has its own gas density sensor which is connected by a data bus (LWL). The arc location sensors are also distributed along the GIL, but allow for longer distances of about 3–5 km. These ALU sensors are connected by a module bus.

The data transferred along the GIL is separated into two data bus systems. The module bus transmits the data from the ALC at the connected disconnecting unit along the line to the interface computer (Int.) in the system control centre. The data bus collects the measuring data from the density and temperature sensor at each disconnecting unit for each gas compartment to the interface computer in the control centre.

5

Environmental Impact

5.1 General

The environmental impact of new transmission lines is of great importance. Public awareness is increasing, and the way through the commissioning process of authorities is getting more difficult. Visual impact, electromagnetic fields, size and duration of construction sites, transportation traffic during construction, impact of thermal soil heat-up, impact of insulating gases, recycling of materials and lifecycle assessment are some topics connected to environmental impact. The GIL offers some good features in these areas.

The public sensitivity to landscape and the protection of natural resources is increasing. In sensitive areas like the mountains (European Alps) or in coastal regions (Wadden Sea, Germany) new overhead lines are rarely accepted by the public and the authorities. Areas of importance for tourism are more and more sensitive towards the erection of new overhead lines. In some cases, even deconstruction of existing overhead lines and replacement by underground systems are under way (e.g., Denmark). For the public, the improvement of the beauty of the landscape by installation of underground transmission systems will be financed by the additional money brought in by tourism to the region. But new transmission lines are needed to overcome bottlenecks in the transmission network. The main reason today is the increase of regenerative energies entering the network. In sensitive areas the GIL offers an economical and ecological solution [101, 154].

5.2 Visual Impact

The tunnel or trench-laid GIL and the directly buried GIL are non-visible to the public once they are installed and in operation underground. The land above the GIL can be used for agriculture. Large trees will not be allowed because of their roots and the need for accessibility in case of a repair. In Figure 5.1 the non-visible GIL is shown in principle using a tunnel with two three-phase systems (centre of graphic) or six single-phase GIL directly buried (left and right of graphic).

Gas-Insulated Transmission Lines (GIL), First Edition. Hermann Koch.
© 2012 John Wiley & Sons, Ltd. Published 2012 by John Wiley & Sons, Ltd.

Figure 5.1   Non-visible GIL. Reproduced by permission of © Siemens AG

5.3   Electromagnetic Fields

5.3.1   General

Electromagnetic fields are a basic technical and physical aspect in high-voltage equipment. They need to be managed technically and physically in the equipment to make it function reliably. The materials chosen must be stable in electromagnetic fields for a long time, for a lifetime of 50 years or more, to provide long-term availability of the expensive high-voltage equipment. This area of electromagnetic fields is constantly covered by the product development of manufacturers and the research work of universities and institutes. Many publications in the various fields document this [155–157]. Here, the focus of electromagnetic fields is on the GIL and its external influences.

5.3.2   Basic Theory

On going into the topic of electromagnetic fields, many keywords and slogans are used: risk, cost, environment, electro smog, failure, emission, immission, future, health, EMC. The list could be longer, if you think about it. To bring about some order, a division can be made into impact on equipment (electromagnetic compatibility, EMC explains the good cooperation of devices and systems) and humans (human impact). What we are really talking about is mathematically expressed by the Maxwell equation:

$$rot\,\underline{H} = (\kappa + j\omega\varepsilon)\underline{E} \tag{5.1}$$

$$rot\,\underline{E} = -j\omega\mu\,\underline{H} \tag{5.2}$$

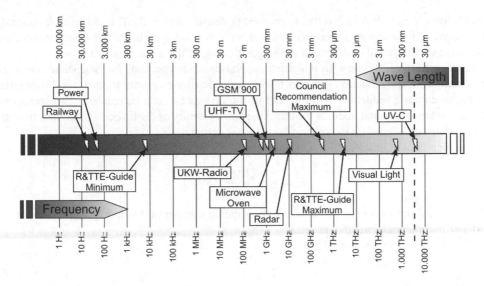

Figure 5.2   Frequencies of technical and natural electromagnetic fields.

The Maxwell equation describes the magnetic field *rot H* and the electric field *rot E* in dependence of frequency ω, the material constants ε and μ and the coupling factor κ. At lower frequencies (e.g., 50 Hz) the magnetic and electrical fields are not coupled, at higher frequencies (e.g., 100 MHz) they are.

In nature we have the magnetic field of the earth, which is constant between the North Pole and the South Pole and has an equal strength of about 40 μT. Electrically constant fields occur in the atmosphere and show values between 20 kV/m in thunderstorms and 0.5 kV/m on a nice day. Electromagnetic fields of high frequency in nature come with lightning strikes (typical frequency 200 MHz), or with light (wavelength 400 nm) or even cosmic radiation (wavelength 100 nm). The technical and natural frequencies of electromagnetic fields are shown in Figure 5.2.

The magnetic field strength H is defined as

$$H = \frac{I}{2 \cdot \pi \cdot r} \; [A/m] \tag{5.3}$$

The magnetic field strength equations describes the magnetic field H depending of the current I, the radius r and the constants 2 and π.

The magnetic flux density B is defined as

$$B = \mu_o \cdot H \; [Vs/m^2] \tag{5.4}$$

The magnetic flux density equation describes the magnetic flux density B depending of the magnetic field H and the permeability constant μ_o.

The unit Vs/m^2 is equivalent to one Tesla (1 T).* To get a feeling for the sizes, the following example is given. A conductor carries a current of 100 A over a distance of 10 m. The magnetic

---

*In some countries the older non-international unit Gauss (G) is also used, where $1\,G = 10^{-4}$ T.

field is then $H = 1.59$ A/m and the magnetic flux density is $B = 2$ μT. In this simple example the magnetic flux density is linearly reduced with the distance. In a three-phase system the reduction is additionally influenced by the positioning of the three conductors.

Because of the 120° phase shift of the three phases, and depending on the distance between the conductors, the reduction is higher than in a single-phase system. In the typical triangular laying of cables a reduction by the square of the distance r^2 can be reached. This triangular laying, on the other hand, reduces the transmission capability of cables because of their thermal limits. One cable heats up the other if laid close together.

The electrical field strength E is defined as

$$E = \frac{Q'}{2 \cdot \pi \cdot \varepsilon \cdot r} \; [V/m] \tag{5.5}$$

The electrical field strength equation describes the charge per unit length Q' depending of the radius r and the dielectrical constant ε.

The charge per unit length Q' is

$$Q' = C' \cdot U \tag{5.6}$$

The charge per unit length Q' equation is depending of the voltage U and the capacitance per unit length C'.

This electrical field strength is only then measurable if no electrical shielding is effective (e.g., overhead lines, radio transmitters). In shielded systems like cable and gas-insulated lines, the outer electric field is zero ($E = 0$). This shielding can be made by cable shields, grounded metallic enclosures (transformers, GIL, GIS) or building steel reinforcements.

To protect humans against the influences of radiation made by technical systems, rules, standards and laws are released by national and international organizations as shown in Figure 5.3.

The sensitivity towards electromagnetic fields in the public is much higher these days than it was some years ago. Only in the last few years have electromagnetic fields become a strong concern, and seems that this concern will get even stronger. Nicola Tesla found in 1880 that high frequencies warm up the human body (first microwave oven). Heinrich Hertz found in 1890 the electromagnetic wave which was mathematically explained by James Clerk Maxwell in 1873. It then took almost 100 years until the first standard was released – VDE 0848 [158] in Germany in – giving maximum values for electric and magnetic fields under certain conditions (only accessible to maintenance people, short time, long time, public). Today we have European and international standards and also national laws. The tendency in all cases is that the maximum allowed values are getting reduced [159].

5.3.3   Maximum Field Values

The situation concerning maximum field values for electric and magnetic fields is that a wide influence comes from physical values as shown in formulae, biological values such as current density in the human body (J [A/m^2]) or the specific absorption rate (SAR [W/kJ]). Also, influences occur on human reflexes, abnormality, dysfunction of nerves or epidemic investigations with statistical evaluations [36, 61, 159].

Limit values are found by defining values which lead to failure or malformation. A safety margin for such values is defined. Then, by taking into account the type of fields and exposure

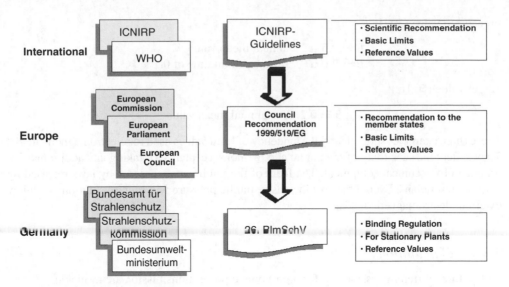

Figure 5.3 Organizations releasing rules and limit values for human protection. Reproduced by permission of © Siemens AG

time, reference values are defined. For example, current density in human bodies: the lowest value leading to failure or negative influences is 100 mA/m². This value is then divided by 10 for short-time exposure with another division by 5 for public exposure, which results in a basic limit value of 2 mA/m².

This systematic now gives a wide field of interpretation. In Germany today the limit values according to the "26. Verordnung zur Durchführung des Bundes-Immissionsschutzgesetzes (26. BImSchV)" [161] are

$$B = 100 \ \mu T(50 \ Hz); \quad E = 5 \ kV/m$$

These values are for overhead lines, cables and substations.

In Switzerland the "Verordnung über den Schutz vor nichtionisierender Strahlung (NISV)" from December 23, 1999 [162] defines

Limit values for the magnetic field (50 Hz):

$$B = 100 \ \mu T$$

Precaution value (50 Hz):

$$B = 1 \ \mu T$$

The precaution value is for buildings with public access and permanent visits, playgrounds for children or free areas with public access.

In Italy, the law for limit values for the protection of professionals and the public from 1999 defines

Limit value (50 Hz):

$$B = 100 \ \mu T$$

Precaution value (50 Hz):

$$B = 0.5 \ \mu T \quad \text{annual mean value}$$
$$B = 2.0 \ \mu T \quad \text{for maximum time of 0.1 s}$$

Target value (50 Hz):

$$B = 0.2 \ \mu T \quad \text{annual mean value}$$

These three examples show clearly the tendency of limit values used for electromagnetic fields. They reflect the sensitivity of the public and the non-acceptance of taking any additional risk produced by technical equipment. The fears of the public are expressed by laws released by nations' politicians. Even if the cost increases, public pressure will require solutions with low external electromagnetic fields.

5.3.4 Calculations

Today, three different possibilities for high-voltage power transmission are available:

- overhead line (OHL)
- cable
- gas-insulated transmission line (GIL).

The most common solution, with tens of thousands of kilometres installed, is the overhead line. Comparing installation costs, overhead lines have the lowest cost. In cases where OHL cannot be built (usually in cities) then cables – or when higher transmission ratings are needed, GIL – have been used.

In the past, electromagnetic fields have played almost no role in the evaluation of the project and the right of way. This has clearly changed today, and electromagnetic fields are now an important factor in transmission projects.

The largest area with high values is generated by overhead lines, as shown in Figure 5.4. The OHL is rated for 400 kV, with two systems of 1000 MVA each. This is equivalent to the rated current of 1450 A. In this case, even at 50 m from the tower the magnetic flux density is $B = 3 \ \mu T$ – too high according to the values allowed in Switzerland and Italy. Some hundred metres are necessary to fulfil the Swiss and Italian requirements.

The highest absolute values are generated by cables. If the cables can be laid close together, then the addition of the three phases helps to reduce the magnetic flux density. The limits for this reduction are given by the thermal layout, not to heat up the cable.

The lowest values are generated by the GIL. The reason for this is the operation as a solid grounded system. The low impedance of the grounded enclosure allows the induction of an inverse current in the enclosure (180° phase shift to the current in the conductor) which is 99% of the conductor current. Less than 1% of the rated current can be measured outside the GIL. The large cross-section of conductor and enclosure and their low impedance produce only low ohmic losses. That means no thermal heat-up.

For transmission projects in the future, magnetic fields will play a more important role. Overhead lines will then be used if the distances beside the OHL are large enough to meet the required values.

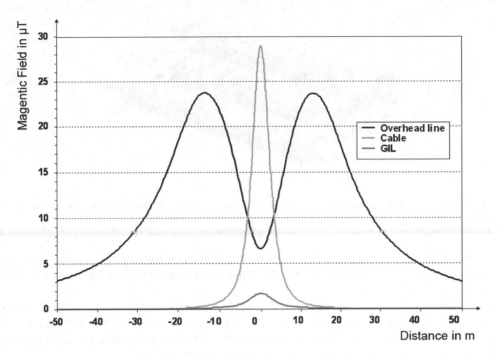

Figure 5.4   Comparison of magnetic flux densities of overhead line, cable and GIL. Reproduced by permission of © Siemens AG

5.3.5   Induced Reverse Enclosure Current

The GIL offers a very specific advantage compared with any other AC transmission technology, and that is the very low magnetic field outside the GIL. The reason is the solidly grounded enclosure and the low impedance of the enclosure pipes. The inductive law induces a voltage in the enclosure which drives a current. The current has almost the same value as the current in the conductor because of the low impedance of the enclosure pipe. The inductive law says that induced current is reversed to its original current and the superposition of both currents results in a very low remaining magnetic field. The electric field is zero because of the solidly grounded enclosure pipe.

In Figure 5.5 the electrical scheme for the induced current is shown.

In a 120° phase shifted three-phase system the induced reverse currents are also 120° phase shifted. Current I_1 in phase 1 induces an enclosure pipe current of phase 1, which induces a current following the inductive law $d\Theta/dt$ that is the reverse of the original current I_1. The circuit of the induced enclosure reverse current is closed by the electrical bridges at both ends A and B. The induced current I_1 circuit is closed by E_2A to E_2B and E_3A to E_3B. For the conductor currents I_2 and I_3 the same rule is valid with 120° phase shift and with the closed circuit of the two other phases, respectively.

The circuit diagram in Figure 5.6 shows, for the conductor current I_1, the loop of the induced reverted enclosure current I_1' with a coupling relationship of 1:1.

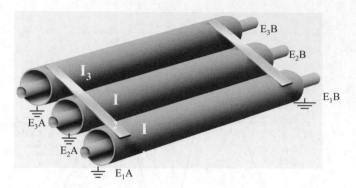

Figure 5.5 Principle of the induced enclosure reverse current. Reproduced by permission of ©
Siemens AG

The parameter K_1 in Figure 5.7 explains the coupling factor between I_1 and I_1' and is chosen
to be 0.99. The parameter $K\varphi$ explains the phase angle coupling factor for φ and φ' and is
chosen to be 0.95 to 0.99.

The high coupling factor of the induced current transformation initiates a reverse current in
the enclosure which has about the same value as the current in the conductor. Both currents
are in phase opposition (180°) and the superposition of both magnetic fields of these currents
lead to the very low magnetic field values shown in Figure 5.4.

5.3.6   EMF Measurements of GIL

5.3.6.1   General

To measure electric and magnetic fields outside the GIL is a standard measuring process and
devices are available. At two installations in Switzerland and Germany the low values of
electromagnetic fields have been verified by measurements [32].

5.3.6.2   GIL Tunnel in Schluchsee, Germany

To prove the level of magnetic fields at the Cavern Power Plant Schluchsee in Germany, field
calculations and measurements have been carried out [32]. As a basis for the calculations, the

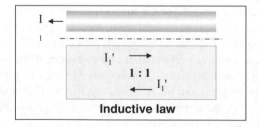

Figure 5.6 Circuit diagram of the induced reverse enclosure current. Reproduced by permission of ©
Siemens AG

$$\textbf{Formula (1):} \quad I_1' = k_1 \bullet I_1$$
$$k_1 = 0.99$$

$$\textbf{Formula (2):} \quad \varphi_1' = k_\varphi \bullet 180°$$
$$k_\varphi = 0.95 - 0.99$$

Figure 5.7 Current and phase angle coupling factor. Reproduced by permission of © Siemens AG

induced currents of the enclosure pipe have been measured. The enclosure pipes are grounded and connected at both ends. The calculated enclosure currents show almost the same value as the current in the conductor and are 180° phase shifted. For the magnetic field outside the GIL the superposition of enclosure and conductor current gives the resulting value. Because of the 180° phase shift the resulting magnetic field is very low at levels of 5% to 9% of the magnetic field of the conductor.

These values have been measured by the use of Rogowski coils and have proven the calculated values. The calculation includes the geometric location of the conductor and enclosure pipe and the current by value and phase angle.

For a symmetric load of 700 A in each three-phase GIL system the calculated magnetic induction in the walkway in the middle of the tunnel is 25 µT. To verify the calculations, magnetic field measurements have been carried out at rated power during the pumping operation of the cavern power plant. To calculate the effective values, standard DIN VDE 0848-1 [158] has been taken. In Figure 5.8 the calculated results of the magnetic induction are shown.

The comparison of the calculated and measured values in the walkway of the tunnel shows differences of less than 20%. Some larger differences between calculated and measured values are shown in the lower area of the tunnel, where auxiliary low-voltage cables are influencing the measurement. Also, the field influences of steel structures in the tunnel are a reason for the differences. Steel has a very high relative magnetic induction. A third reason for the differences are induced currents in cable shields, grounding wires and connected steel structures. All these influences of the real tunnel are very complex to simulate in the calculation and are the reason for the differences between calculated and measured values. In conclusion, it can be said that the allowed limiting value of 100 µT at the walkway in the middle of the tunnel has been obtained by the factor 4 with 25 µT.

The measurements in the tunnel were carried out during a very constant current rating when the hydropower plant was used in the pumping mode. A constant value of 700 A was then transmitted by the GIL.

To measure the current a Rogowski coil was used, which was laid directly around the GIL enclosure. This increased the accuracy of the measurement because the Rogowski coil measures any current going through the measuring ring.

The magnetic flux density was measured in the tunnel in a 20 cm × 30 cm grid. The magnetic sensor measures in three dimensions by three orthogonally placed coils. The values in Figure 5.9 are r.m.s. values of the x, y and z components of the magnetic flux density. The

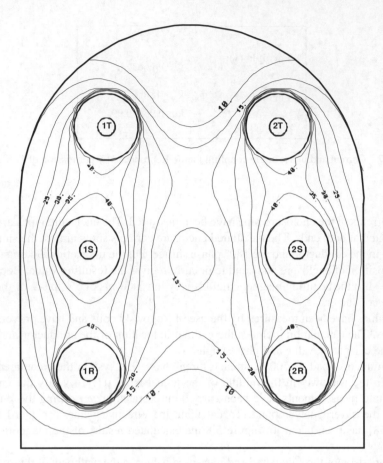

Figure 5.8 Calculated results of the magnetic induction. Reproduced by permission of © Siemens AG

calculations use the formula

$$B_{eff} = \sqrt{B_{X,eff}^2 + B_{Y,eff}^2 + B_{Z,eff}^2} \qquad (5.7)$$

The magnetic flux density was measured in the tunnel and also between the GIL pipes. The highest value in the tunnel walkway is 20 μT [32].

Comparing Measurements and Calculations   In Tables 5.1 and 5.2, comparisons of the measured and calculated values are shown.

The measured and calculated values of the current measurement with the Rogowski coil are very close together, as shown in the comparison of Figure 5.10.

The differences are related to influences coming from other currents in the tunnel, as there was a 6 kV cable under the walkway and a 400 V cable on an installation under the roof. Related to the 700 A of the GIL, these differences are small. The same size of differences was also seen in comparison of the measured and calculated magnetic flux density. The tolerances are below 10%.

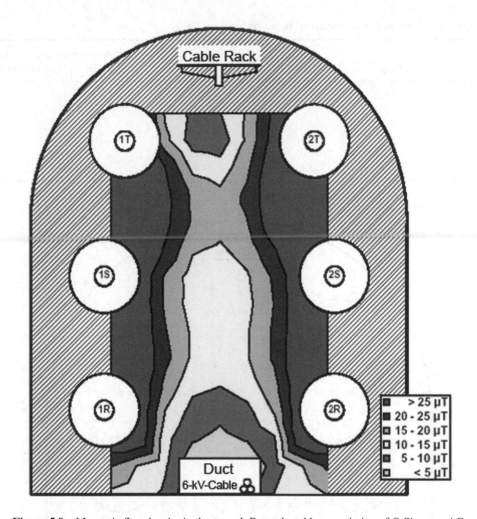

Figure 5.9 Magnetic flux density in the tunnel. Reproduced by permission of © Siemens AG

Table 5.1 Comparison of calculated and measured values at 700 A

System	Phase	Calculated (A)	Measured (A)	Difference (A)
1	1	40.2	49.4	+9.2
	2	61.0	60.3	−0.7
	3	40.4	38.9	−1.5
2	1	40.2	50.1	+9.9
	2	61.0	61.5	+0.5
	3	40.4	50.8	+10.4

Table 5.2 Calculated induced currents in the enclosure at 700 A

System	Phase	Value (A)	Phase angle (°)	Difference value between enclosure and conductor (A)
1	1	685.2	−176.91	40.2
	2	699.3	64.99	61.0
	3	713.1	−56.90	40.4
2	1	685.2	−176.91	40.2
	2	699.3	64.99	61.0
	3	713.1	−56.90	40.4

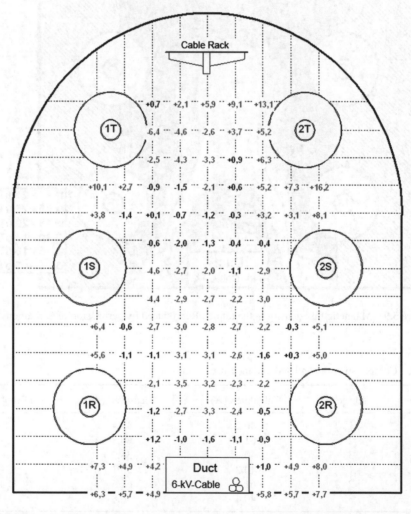

Figure 5.10 Comparison of calculated and measured values at 700 A. Reproduced by permission of © Siemens AG

Discussion and Results The comparison of the measured values with today's required values of max. 100 µT for magnetic flux density shows that even in the tunnel, it is met by the GIL application [159]. At distances of less than 10 m away from the buried or tunnel-laid GIL, even the very low requirements valid today in Switzerland (1 µT) [36] and in Italy (0.3 µT) can be met using GIL at a current rating of 3150 A.

The measurements carried out in the tunnel with the GIL also showed that the calculations made so far with the NETOMAC programme [121] have high accuracy. The differences are below 10%. This means that the coupling between the conductor and the enclosure is correctly modelled. It also proves that GIL lines of some 100 m length have a good coupling between conductor and enclosure. The low impedance in the coupled circuit allows a high reverse current in the enclosure. The resulting magnetic field outside the tunnel, and over a few metres beside a directly buried trench, do have values below 0.3 µT as required in Italy and Switzerland for new installations [157].

5.3.6.3 GIL Tunnel at Geneva, Switzerland

The GIL application at Geneva Airport has very low magnetic fields, as per requirements because of the closeness to the airport and fair exposition halls at the airport's PALEXPO complex. The maximum magnetic flux density above the GIL tunnel at the floor of the building, at about 5 m distance, was only 1 µT. This value has been measured and calculated at the rated voltage and current and is shown in Figure 5.11.

Electric Field The enclosure pipe is solidly grounded and therefore zero. The solid grounding is made at each steel structure in the tunnel and connected to the tunnel grounding system. This grounding is effective for power frequency voltage and also for high-frequency, transient voltage.

The measurements at PALEXPO, Geneva were carried out with two GIL systems under operation at a current of 2×190 A. Based on the measured values, the magnetic induction was calculated for the load of 2×1000 A. Inside the tunnel between the two GIL systems the maximum magnetic induction amounts to 50 µT.

The magnetic field at right angles to the GIL tunnel is presented in Figure 5.12. The measurements have been taken 1 m and 5 m above the tunnel, which is equivalent to street level and to the floor of the exhibition hall.

The magnetic induction on the floor of the fair building is relevant for fulfilling the Swiss regulations for continuous exposure to magnetic fields. The 1 m maximum value amounts to 5.2 µT above the centre of the tunnel. The maximum induction at 5 m above the tunnel is 0.25 µT. This result is only 0.25% of the permissible German limit [158] and 25% of the new Swiss limit.

The calculation and measurements also show that it is not sufficient to look only to the GIL. The complete 220 kV transmission system has impact on the magnetic field in the tunnel. At symmetrical load the ground wire in the tunnel is carrying an induced current from the GIL. This ground wire current has some impact on the magnetic field in the tunnel and – including the connected cable shielding, ground wires and steel structures – local higher magnetic field

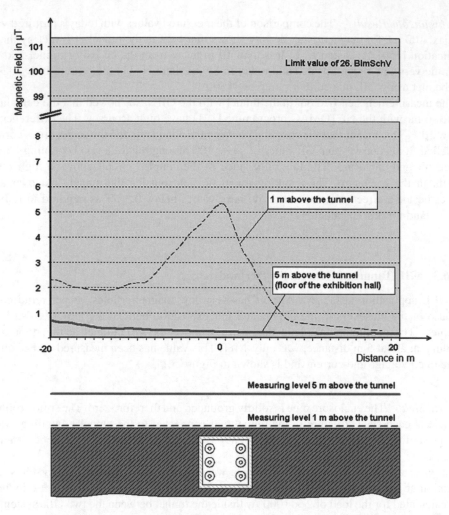

Figure 5.11 Measured flux density of the GIL at Geneva Airport above the tunnel. Reproduced by permission of © Siemens AG

values can be measured. For this reason the magnetic flux density on the left-hand side of the tunnel is increasing at distances of about 20 m, see Figure 5.12.

In total, the measured magnetic field on top of the tunnel is low and fulfils the requirements of the operator and the Swiss laws of less than 1 µT. The results also show that for the system configuration the grounding wires, parallel auxiliary cables and civil structure – including steel structures and steel reinforced concrete – need to be taken into account.

The magnetic inductions above the GIL trench are negligible and meet the Swiss requirements under full load conditions. The results show that for the design of the system the magnetic fields induced by the grounding system also need to be considered. All measured and calculated values of the induction from the GIL are far smaller than those for comparable overhead lines and conventional cable systems [32, 159].

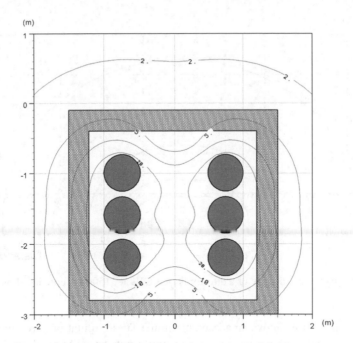

Figure 5.12 Measured values of the magnetic induction above the GIL tunnel at PALEXPO, Geneva, Switzerland, at a rated current of 2 × 1000 A. Reproduced by permission of © Siemens AG

5.3.7   Direct Buried GIL

For directly buried GIL the principle of calculating the magnetic field around the GIL is the same as above. The distance to the GIL is getting short because the GIL is laid about 1 m under ground. In Figure 5.13 the calculation result for a 420 kV GIL at 3150 A current is shown.

The low magnetic field of the GIL shows less than 10 µT right on top of the buried GIL at 1 m distance. At less than 10 m distance the magnetic flux density is down to 0.3 µT at a current of 3150 A [32, 159].

5.4   Gas Handling

From an environmental point of view gas handling is an important aspect. Using SF_6 as an insulating gas with a high global warming potential it is necessary to have a closed-loop process at every step of gas handling. From the first filling, over maintenance and dismantling at the end of the lifetime the insulating gas needs to be kept in compartments as explained in Section 3.9. This closed-loop gas-handling process is well developed and IEC gas-handling standard 62271-303 and IEC 62271-4 [51, 58] gives detailed information and instructions. The closed loop gas handling concept is covering the complete life cycle of the equipment and is part of the life cycle assessment (LCA).

5.5   Thermal Aspects

GIL installations buried in a tunnel, or directly buried, produce heat when electric power is transmitted. For project planning it is necessary to study the heat transfer for each project

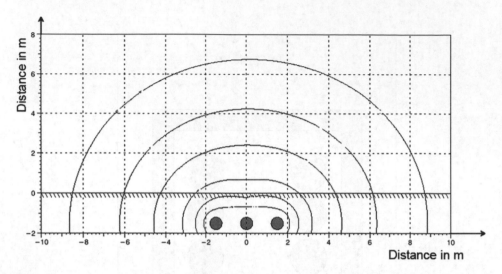

Figure 5.13 Magnetic flux density of a directly buried GIL at 3150 A. Reproduced by permission of © Siemens AG

to avoid temperature rises above an acceptable limit. These limitations are given by the soil requirements to avoid negative impacts, for example non-reversible soil dry-out or impact on vegetation. See also the thermal design criteria given in Section 3.7 for the different laying options.

5.6 Recycling

For good ecological acceptance, recycling is an important aspect. The GIL is easy to dismantle at the end of its lifetime. The different materials – like aluminium pipes, insulators or gas – are all separated by design and can be recycled.

The main components of a GIL are:

- aluminium pipes
- N_2 and SF_6 insulating gas
- cast resin epoxy
- aluminium cast enclosures
- corrosion coating (for buried GIL)
- steel structures in tunnels and shafts.

The main volume of materials used comes from aluminium pipes of the enclosure and conductor (more than 90%). The enclosure pipe is an AlMg alloy and the conductor pipe is made of pure aluminium (99.5% Al), called electrical aluminium. The enclosures of the disconnecting unit and angle unit are moulded aluminium of AlSiMg alloys. The enclosure of the compensator unit can be made of aluminium or steel bellows. The recycling of these valuable raw materials is made by aluminium pipe of sheet manufacturers.

The insulators are made of epoxy cast resin and can easily be separated with dismantling. The recycling of cast resin will not be with new insulators, because of the high purity of the material requirements; it will be recycled in other plastic processes.

Recycling of the insulating gases is done by storing the gases in compartments and sending the gases back to the manufacturer to be used in the process for new gases.

When the enclosure pipe is coated with a passive corrosion protection the coating will be taken off the aluminium pipe in a factory process. This can be the same factory as the one that applied the coating.

Steel structures are dismantled and the steel is recycled in the steel production process.

In total, the recycling of a GIL is an easy process because the different materials can be separated for the different recycling processes.

5.7 Lifecycle Assessment

From an ecological point of view the installation of a GIL over long distances needs to be evaluated in a lifecycle assessment. The high installation cost and the low transmission losses need to be averaged over the long period of using the installation and the ability to transport high amounts of energy over long distances [202].

The impact on the ecological system occurs only once, and then for 50 years and more no access can be expected. One of the oldest GIL was installed in 1974 in the Black Forest in Germany and is still in full operation without any sign of ageing, and this after more than 36 years.

The principle of lifecycle assessment is the view of all impacts coming from the system for manufacturing, operation and dismantling. This comparison is usually normalized to the CO_2 footprint. This means that any energy used to produce the raw materials, to operate the transmission line and to dismantle and recycle the equipment. This CO_2 footprint is explained in detail in Section 5.8.

5.8 CO_2 Footprint

The CO_2 footprint is a possible evaluation method to compare and evaluate technology concerning its global warming contribution. GIL has a great potential to improve the CO_2 footprint position for the transmission of high electric power ratings over long distances because of its low transmission losses. At high current ratings of 3150 A and a voltage level of 400 kV, the power losses are less than 0.01% per kilometre. For an 800 kV transmission line the losses are reduced to one-quarter of the 400 kV value at the same power transmission rating. Transmission losses are the main contribution to the CO_2 footprint [202].

The higher the voltage level the lower the losses per transmitted power. For long distances, 800 kV or even 1000 kV are used. To connect the large regenerative energy resources which are far away from load centres (wind in the North, sun in the South), UHV plays a more important role and GIL offers a solution. Not only the low losses of the GIL transmission line will contribute to the low CO_2 footprint, but also the fact that regenerative energy resources can be connected to the transmission network and will contribute to reduce the overall CO_2 footprint.

A CO_2 study for GIL has been carried out and produced the following results [202].

In accordance with ISO 14040 [199], a lifecycle assessment and evaluation has been carried out. The lifecycle inventory analysis (LCI) identifies all materials and energy-consuming processes involved for the lifetime of the product. The goal of the study was to identify the CO_2 impact of a long-distance GIL over its total lifetime, including the construction and manufacturing process. Even the possibility of a failure in the GIL gas-tightness, and losing

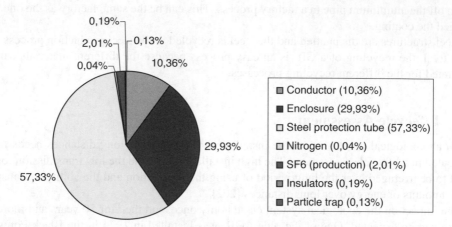

Figure 5.14 Distribution of CO_2 contribution: 1000 MVA GIL. Reproduced by permission of ©
Siemens AG

SF_6 to the atmosphere, was investigated. The results of the distribution of the CO_2 contribution
of a 1000 MVA, 2000 MVA and 3000 MVA GIL are shown in the following figures.

Figures 5.14–5.16 show the result for the equipment of a GIL with 500 mm enclosure
diameter once for 1000 MVA, once for 2000 MVA and once for 3000 MVA transmitted power.

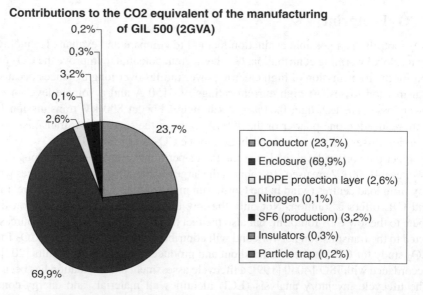

Figure 5.15 Distribution of CO_2 contribution: 2000 MVA GIL. Reproduced by permission of ©
Siemens AG

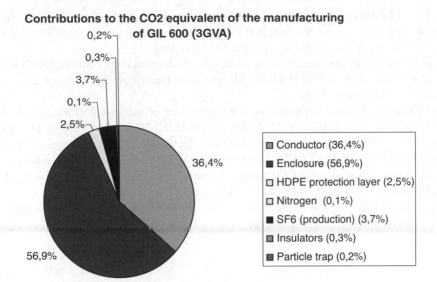

Contributions to the CO2 equivalent of the manufacturing of GIL 600 (3GVA)

Legend:
- Conductor (36,4%)
- Enclosure (56,9%)
- HDPE protection layer (2,5%)
- Nitrogen (0,1%)
- SF6 (production) (3,7%)
- Insulators (0,3%)
- Particle trap (0,2%)

Figure 5.16 Distribution of CO_2 contribution: 3000 MVA GIL. Reproduced by permission of ©
Siemens AG

In both cases the production of the enclosure pipes is a dominating value of the CO_2 footprint,
with 67.2% and 69.9%. When the values of the conductor pipes of 23.4 and 23.7 are added,
then more than 90% of the CO_2 footprint is covered by the aluminium pipes.

In the case of the 600 mm pipe diameter these values are 56.9% for the enclosure and 36.4%
for the conductor, also above 90%.

Type of CO2 contribution	1 000 MVA	2000 MVA	3000 MVA
Manufacturing	3,78%	2,00%	1,52%
Electrical losses	89,60%	95,50%	96,31%
SF_6 leaks	6,62%	2,50%	2,17%

Figure 5.17 Total contribution of CO_2 footprint for 1000, 2000 and 3000 MVA GIL. Reproduced by
permission of © Siemens AG

The SF_6 contribution is surprisingly low, with values between 4.5% for the 500 mm 1000 MVA GIL and 3.2% for the 800 mm 2000 MVA version. Insulators are at 0.3% of the CO_2 footprint, with a very low contribution to the CO_2 footprint.

In comparison with the operational transmission losses and their contribution to the CO_2 footprint of 1000, 2000 and 3000 MVA GIL, the manufacturing-related contribution is very small (as shown in Figure 5.17).

The higher the transmission value the lower the share of the manufacturing, from 3.78% for 1000 MVA to over 2.00% for 2000 MVA down to 1.52% for 3000 MVA. The absolute majority of the CO_2 footprint is coming from the electrical transmission losses, with 89.6% for 1000 MVA to 95.5% for 2000 MVA and 96.3% for 3000 MVA transmitted power.

Again, the contribution to the CO_2 footprint coming from SF_6 leaks is surprisingly low, at only 6.6% for 1000 MVA, 2.5% for 2000 MVA and 2.17% for 3000 MVA.

In conclusion it can be stated: the lowest CO_2 contribution is reached by the transmission system with the lowest transmission losses.

6

Economic Aspects

6.1 General

Gas-insulated transmission lines are long-living investments at a high installation cost level. They are long-term investments and stay in service for decades. The first GIL was installed in 1974 at the 400 kV transmission voltage level in a pumped hydro storage plant and still does its duty now, more than 36 years later, without any substantial maintenance required since the beginning. And many more projects have been built and are in operation since this time world-wide with the same positive operational experiences.

With GIL, the first-place investment cost is high for the product and its installation. The operation cost (mainly the losses from power transmission) is low. The high power transmission capability generates high income on a power transmission rating per kilowatt hour. So the GIL can be an investment for the future and a possibility to generate long-term revenues. The main aspect of why GIL is a good investment is that large-scale regenerative energy resources can be connected to the load centres by underground installation.

In this chapter the different economical aspects are explained and information is given to reduce costs for certain aspects [220].

6.2 Material Cost

The installation cost of the GIL is a large investment in a strong power transmission system. The main part of the installation cost is due to the aluminium pipes for enclosure and conductor. The cost share of the insulators is small, due to the small number needed. Less than 0.5% of the total volume of the GIL is related to insulators.

Disconnecting units and angle units are needed at large distances of 1–1.5 km along the line. Therefore, these materials contribute only a small cost share to the total cost. In total, the cost share of materials is around 40% of the project cost.

The second largest cost share comes from the laying and installation process: on-site pre-assembly, on-site transportation, laying into the trench or pulling into a tunnel, welding the joints, fixing on steel structures, on-site high-voltage testing and commissioning. The total cost share is about 40% of the project.

Gas-Insulated Transmission Lines (GIL), First Edition. Hermann Koch.
© 2012 John Wiley & Sons, Ltd. Published 2012 by John Wiley & Sons, Ltd.

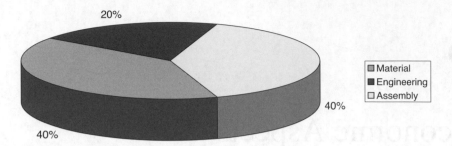

Figure 6.1 Split of cost share of GIL. Reproduced by permission © Siemens AG

The smallest part of the cost is in engineering, planning, calculation, project management and all kinds of studies needed before, during and after the installation of a GIL. This cost share of the project is about 20%. An overview of the cost share is shown in Figure 6.1.

When looking in more detail at the material cost, the aluminium pipes show the largest share of the cost. The reason for this is simple: the volume of aluminium needed and the quality of the manufacturing process.

The volume of aluminium contributes only by the kilogram of aluminium needed. The weight of the aluminium pipes is defined by the diameter of the pipe and the wall thickness of the pipe. The diameter of the pipe is defined by the rated voltage for the enclosure and the conductor pipe, as explained in Section 3.2. The wall thickness is the parameter influencing the current-carrying capability for optimization of the specific project requirements. The pipe diameter is the parameter for voltage levels, e.g. 420 kV has 500 mm and 800 kV has 630 mm diameter. Depending on the power to be transmitted per system, 3000 MW for the 420 kV GIL or 4000 MW for the 800 kV GIL, the volume of aluminium will be optimized for the transmission task.

The aluminium pipes are produced under the requirements of the high-voltage design where tolerances of diameter variations and limits of the ovalization of the round pipes define the manufacturing process. Today, the quantity of GIL pipes per year is relatively low. Some tens of kilometres of GIL pipes are ordered each year, with an increase year on year over the last 10 years. The consequence of this relatively low amount of aluminium pipes is that almost all pipes for GIL are produced by one aluminium pipe manufacturer. The maximum capacity of one spiral welding machine is about 120 km of pipes per year. That means the machine is only used at 10%, a low value in production and free capacities can be used. This will reduce the pipe production cost.

It is expected that (if the capacity is fully used) the cost for aluminium enclosure and conductor pipe manufacturing will go down. This will bring a major cost reduction for the total cost as the aluminium pipes have a large cost share of a GIL project. It can be expected that, for the manufacturing process, a cost reduction of 30–50% is possible.

Insulators for GIL will be needed in larger numbers when more projects come to fruition. Their cost impact is low and there is also the cost reduction resulting from higher quantity. The insulators are produced with the same production equipment as those for GIS and therefore the production factor can be optimized jointly for GIL and GIS insulators. The total available capacity for insulators is large and can easily be extended by copying existing production units, if needed.

Insulating gases used for GIL are sulphur hexafluoride (SF_6) and nitrogen (N_2). A typical gas mixture ratio is 80% N_2 and 20% SF_6. N_2 has relatively low cost and is widely available compared to SF_6. SF_6 is relatively expensive and only gets produced at the quality required for high-voltage applications in a few facilities around the world. SF_6 and N_2 are available in large quantities and the cost can be reduced per kilogram if large amounts are ordered for GIL projects.

The large percentage of N_2 in the gas mixture keeps the total cost low and insulating gas cost forms only a small part of the total cost. N_2 is available at low cost world-wide.

Disconnecting units, compensator units and angle units are relatively expensive as they are made of cast aluminium housing or use special steel or aluminium alloy for compensation bellows. The use of these units is strongly dependent on the project requirements and the type of laying (directly buried or in a tunnel). The more bending and directional changes are required the higher the project cost, and this is really a cost driver.

For standard routings directly buried without angle and compensator units, and disconnecting units at 1000 to 1500 m distance, the cost for these disconnecting units is negligibly low. If shorter distances for disconnecting unit, angle unit for directional changes and compensator unit are needed these units might dominate. The project routing has a big influence on the cost. Disconnecting units, angle units and compensator units can really be cost driven in specific projects. Here, project planning is very important to develop alternative solutions.

6.3 Assembly Cost

The principle of assembling GIL is the same as pipelines: transportable lengths of pipes are jointed on-site. The cost of assembling the GIL is very much dependent on the conditions at site: the maximum transport length of pipes, the number of angle units, the site conditions for handling the equipment and the working conditions and rules for the workers. All these parameters are different for each project and need detailed project planning to find the optimum assembly process at the lowest cost. The following are some key points.

Maximum Transport Length   The maximum transport length for the enclosure and conductor pipes is defined by the restrictions of road or railway transportation. The longer the transport length the less the joint works and the lower the assembly cost. Each additional metre length of the pipes will reduce the number of on-site weldings. If 10 m is the limit then 100 weldings are needed for 1 km, if 20 m is possible only 50 weldings are needed. For three phases the total number of 300 weldings for 10 m pipes goes down to 150 m for 20 m pipes. One joint of enclosure and conductor, including quality testing, will take 3–4 hours for two to three workers. This makes it clear that the maximum transport length has a strong impact on the project cost and can be cost driving.

Elastic Bending   The use of the elastic bending of the GIL with a bending radius down to 400 m should be the first choice when the routing needs to be adapted to the landscape. In pipeline experiences under normal site conditions (not mountains, swamps or rocks), one angle unit is needed to change the direction more than the 400 m bending radius can offer. The use of elastic bending is cost reducing in a project.

The angle units are more costly than the enclosure and conductor pipe. The angle unit enclosure is made of cast aluminium and in case of directly buried GIL needs to be strong enough to take the forces of thermal expansion. Therefore, high wall thickness is needed. To connect the angle units to the enclosure pipe two welding joints are needed. One will be done in the workshop for pre-assembly, the second will be done on-site with the laying in the trench. It is clear that with each angle unit the cost per kilometre goes up and if too many angle units are used the cost might be too high.

Site Conditions for Handling   The works on-site dominate the laying speed and with this the total project cost. There is an optimum of project time versus the need for equipment and workers. This optimum needs to be found for each project because the laying conditions on-site are different.

In principle, the materials for pipes, insulators, contact systems, disconnecting units and angle units are delivered directly to the pre-assembly on-site. Close to the trench or tunnel the materials are pre-assembled to laying units, which are then transported to the trench or tunnel as explained in Section 3.7. The way the site conditions are solved in the project plan has a strong impact on the total cost. Here, close cooperation with the user is needed.

Working Conditions and Rules for Workers   To work on-site the conditions and rules for workers are given in local regulations and laws which need to be followed. Safety rules, working hours, shift restrictions and health regulations have an impact on the work organization for the project. This will have an influence on the total time needed for the project and with this a strong influence on the total cost.

The assembly cost is strongly related to each project and needs to be developed together with local authorities. For each project the work plan will be different and therefore the cost.

6.4   Transmission Losses

Transmission losses are a major factor of the operation cost, see also Section 4.3. The thermal losses are produced by the current in the GIL for the lifetime of the equipment, which is more than 50 years. In the following, a comparison of GIL losses versus overhead line losses is given to show the advantage of GIL.

In Table 6.1 the cost differences of transmission losses are calculated for the overhead line and GIL. The values are taken from the tables of parameters in Section 4.3 for 2000 MVA transmission power, which is equivalent to about 3000 A. The difference in losses between GIL and overhead lines for a 100 kilometre system length is 64 MW. To evaluate these losses, an energy price of 5 euro cent per kWh is taken. This sums to costs for yearly losses of 27 520 000 euro (about 10% of the investment cost). This calculation only takes into account the ohm's losses of transmission. Other aspects of low maintenance cost and high reliability are not counted. Also, the possibility that the required permits to build a long-distance transmission line are easier to get for an underground system.

The transmission losses can add a major cost advantage to the transmission system and high amounts of money can be saved. Or, if seen the other way around, the saved transmission losses can be sold at the end of the transmission line to consumers [132]. See also Section 5.8 for the CO_2 footprint.

Table 6.1   Cost differences of transmission losses of OHL and GIL

		OHL	GIL
Transmission power	MW	2000	2000
Losses per system metre	W/m	820	180
Losses of 100 system kilometre	MW	82	18
Difference between GIL and OHL	MW	Δ 64	
Difference cost of losses per year (Energy price 5 Ct/kWh × 8600 h × 64 000 kW)		€ 27 520.000	
to compare: investment GIL, 100 km		€ 300 000.000	

6.5   Cost Drivers

The transmission system of the future needs to cover long distances for regenerative energy connection to consumers in the load centres. The price to build such a transmission system will play a key role in how fast this new network will develop. Prices and cost need to be seen in a total lifecycle. Installation cost, transmission losses, financing cost, cost for long duration time of permissions and cost for acceptance by the public are the main cost elements to be evaluated.

The other point is that a future transmission network needs to transmit a high amount of power (several tens of Gigawatts) over long distances (several hundred kilometres), which can only be reached when solutions are thought big. An overall cost reduction can only be reached if large initial investments into the transmission network are made. Most distances may be solved by 400 kV or 800 kV or even 1000 kV transmission overhead lines. But it can be seen that longer sections need to be undergrounded. In this case the GIL offers a solution at a higher price than overhead lines. The total cost is a mean value of overhead line and GIL.

The advantage of using an underground transmission with GIL is that in areas where commissioning problems with the public would delay the construction, and with this the regenerative electric energy would not be connected to the transmission network and the consumers, the total may increase due to an inefficient power generation mix and reducing the generation of installed regenerative energies (e.g., wind farms).

The large regenerative power generation resources in wind energy on land or offshore (e.g., the North Sea potential alone is estimated to be 300 GW) and solar energy from the deserts (the solar potential of a 500 km by 500 km square in the Sahara could produce the world's electrical energy consumption of today) are available today. The investments into regenerative energies are increasing strongly world-wide, with ever-increasing wind farms (on land and offshore today 500 MVA are under construction and 1000 MVA are in planning) and solar parks (500 MV parks are now under construction).

The missing final link to bring these energies from the generation location to the consumers in densely populated areas, industrial countries, metropolitan areas, merging countries and soon to be developing countries are the transmission line capabilities. As in China today, high-power, long-distance transmission lines are needed to bring 3, 5, 8 or 10 Gigawatt per transmission system to the consumer areas.

The initial investment in GIL is used to solve the underground sections at higher costs compared with overhead lines. But GIL can make a transmission project possible and if not, the regenerative energy source cannot be used. This has to be taken into account if costs, mainly initial investment costs, are evaluated.

It is similar to the development of the railway system world-wide. In any country with strong national investment plans the railway system developed very well – starting in the UK, Germany, all over Europe, the USA and later Japan, the infrastructure is well developed today. The same can be said for the highway system. The initial investment cost of railways or highways was supported by governments and brought benefit to all nations and the public. A new overlay transmission network, a "Supergrid" can have the same trigger effect in using regenerative energy sources which are not connected to the network today. This needs to be evaluated when cost drivers and total costs are determined. Nevertheless, it is necessary to identify cost drivers of GIL to allow cost-optimized solutions. The following cost drivers for GIL can be identified.

Power Rating   One cost-driving factor is the required power transmission capability.

The overhead line has the lowest investment cost, while the investment costs of GIL and cables are higher.

For the cables, a cost increase is related to the need for forced cooling and duplication or triplication of cable systems.

The GIL cost increase is directly related to the weight of aluminium needed for increasing wall thickness of the conductor and enclosure pipes [132].

One first method to increase the power transmission capacity is to increase the high-voltage rating. The GIL is in operation up to 800 kV. A 2000 MVA GIL system at 400 kV can easily be increased to 4000 MVA by doubling the voltage.

Route Planning   For on-site assembly work the impact of obstacles (road crossings, trenches in rock, river, lakes, etc.) are large cost drivers of a transmission project. During the planning status the route planning needs to be optimized for reduced investment cost.

Site Accessibility   To assemble the GIL the construction site needs to be accessible by truck traffic to deliver the enclosure and conductor pipe. These are not heavy, but are large in volume and length. The continuous work and laying process also needs some space to store pipes and other materials. For the pre-assembly a tent is required close to the site for preparing the GIL sections to be laid.

Spiral Welded Pipe Manufacture   For large transmission projects (e.g., 100 km length) it will be of cost advantage if the pipe production can be located close to the site. The spiral welding machine only needs about 20 m by 40 m in a tent to produce about 120 km spiral welded pipes per year. The pipe material is delivered in coils, which are much more compact to transport. This possibility offers a chance of project cost reduction.

Worker Knowledge   The assembly of GIL needs worker skills which are world-wide and locally available. After some weeks of training, skilled workers are available in large numbers to work in shifts during construction. The total time for construction has a major impact on the total project cost.

7

Applications

7.1 General

In this chapter some typical applications are shown as they have been installed world-wide over the last four decades. Starting at the Massachusetts Institute of Technology in the 1960s, gas-insulated technology has been a success world-wide at different voltage levels, in different fields of applications and climatic conditions. Some hundred kilometres phase length of GIL are installed world-wide, as Table 7.1 shows.

The reason for applying GIL in the transmission network is related to dense population, environmental protection and conjunction with infrastructure.

Dense Population   In densely populated areas only limited space is available, the cost of ground is high and public opposition is strong. In such cases overhead lines may not be built and the GIL is used in some cases or cables in other cases.

Environmental Protection   In environmentally protected areas the use of overhead lines is difficult to get permission for. In some cases an underground solution with GIL or cable will be chosen.

Infrastructure   The use of GIL is mostly in conjunction with infrastructure projects, e.g. hydropower dams, nuclear power plants, gas or oil power plants, coal power plants. In such infrastructure projects – which include substations – GIL is used to connect several parts of the high-voltage system inside such infrastructure facilities.

Future fields of application are developing together with infrastructure projects in traffic. Such infrastructure projects are railroads, highways or shipping canals which can be combined with high-capacity power links using GIL. The integer safety of GIL that – even in case of an internal failure of the GIL – no danger is posed to its surroundings, allows GIL to be integrated in railway or road tunnels or bridges or beside the tracks and highways [136, 154].

Gas-Insulated Transmission Lines (GIL), First Edition. Hermann Koch.
© 2012 John Wiley & Sons, Ltd. Published 2012 by John Wiley & Sons, Ltd.

Table 7.1 Installed GIL world-wide, status as at 2010

Voltage level (kV)	Length per phase (km)
72/145/172	38
245/300	33
362	15
420	110
550	90
800	3
1200	1
Total	**290**

Basically three different methods of laying GIL are available. Laid directly in the ground, laid in a tunnel or above ground on structures. The laying method above ground on structures is usually used for applications within substations or power plants, where no public access is allowed. If the GIL has to pass through publicly accessible areas when transmitting electrical power at high voltage levels then it must be non-accessible to the public and therefore laid underground, in a tunnel or directly buried.

The directly buried GIL is the most economical and fastest method of laying across the countryside, where accessibility to site works is given. With the adaptations of pipeline laying technology, the ways and methods of laying are similar to those of gas or oil pipelines. The tunnel laying technique is mainly used in metropolitan areas, where accessibility to site works is limited because of streets, houses, etc. If more accessibility is given, tunnels can also be built using prefabricated concrete segments laid in a trench with a final soil coverage of 1–2 m. In all cases, only standard units as described in Chapter 1 are used.

For all laying methods of GIL the elastic bending of the aluminium pipes can be applied to follow a bending radius as low as 400 m. This is basically sufficient to follow the landscape and to avoid the use of expensive angle elements for directional changes. To separate a transmission line into convenient gas compartment sizes which can easily be handled, disconnecting units with gas-tight conical insulators are placed at distances of about 1.0 to 1.5 km along the transmission line.

In the following, some examples are given of how GIL have been used.

7.2 Examples

7.2.1 Schluchsee, Germany, 1975

The first application of a 400 kV GIL was the installation at the Schluchsee hydropower pumping storage plant in Germany. The GIL connects the high-voltage power transformer in the cavern at the mountain to the overhead line on top of the mountain. The GIL went into service in 1975. The system length of the GIL is about 700 m [147].

Figure 7.1 shows a view into the tunnel with two GIL systems fixed to the tunnel side walls. The GIL is installed in a sloop of about 40°. In the middle, a stairway is installed to allow safe walking. On the roof of the tunnel the control wiring is fixed and the thermal expansion bellows are shown in the middle of the photo.

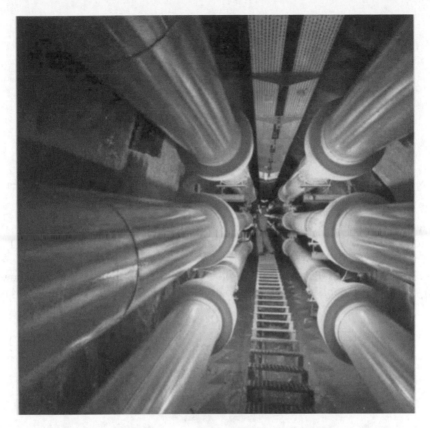

Figure 7.1 Tunnel view of the GIL at Schluchsee, Germany, 1975. Reproduced by permission of ©
Siemens AG

The cavern power plant is part of the 400 kV transmission network in Southern Germany
and delivers on a daily basis peak load energy into the network as required. Usually this peak
load is delivered during the day time and at night the generators operate in the pumping mode
and pump the water from the lower to the upper water reservoir. The first 400 kV transmission
from the cavern to the overhead line was made by oil cable. This oil cable failed and the related
fire destroyed the cable and the tunnel. For the new tunnel a non-flammable GIL was chosen
for the connection.

The cavern power plant delivers 1000 MW peak power into the 400 kV network. The
installed GIL was the first installation world-wide – built in 1974 and in operation since 1975.
The two GIL systems are dimensioned for a maximum operating voltage of 400 kV and a
maximum current rating of 2500 A. The GIL is connected to two machine transformers of
600 MVA each. These transformers connect the 21 kV generator voltage to the 400 kV GIL
and network voltage. In Figure 7.2 an overview of the tunnel cross section is given.

The GIL technology at Schluchsee uses welded joints with sleeves. This jointing technology
connects the single pipes by two orbital welds in a V shape at the sleeve. The enclosure pipes
have longitudinal welding seams. See Figure 7.3.

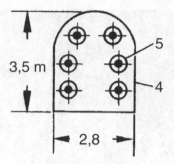

Figure 7.2   Cross section of the schluchsee tunnel. Reproduced by permission of © Siemens AG

The technical data of the Schluchsee GIL is shown in Table 7.2.

The total length of the GIL in vertical and slope section is 700 m for the tunnel, which makes a pipe length of about 4200 m.

The experiences with GIL at the cavern power plant have been very good, for now more than 35 years of operation. There has been no problem in operation. The gas-tightness is high,

Figure 7.3   View into the tunnel with two GIL systems. Reproduced by permission of © Siemens AG

Table 7.2   Technical data Schluchsee, 1975.
Reproduced by permission of © Siemens AG

Rated voltage U_r	420 kV
Rated current I_r	2000 A
Rated impulse withstand voltage U_{BIL}	1640 kV
Rated short-time current I_s	53 kA/1 s
Length per phase	4.2 km
Insulating gas	100% SF_6

so that even after 35 years no insulating gas needs to be added. The welding technology used for the first time in this project has proven to give a reliable and gas-tight connection of the enclosure pipes and the electrical conductivity for the conductor pipe is excellent.

To protect the GIL at the connection to the overhead line, overvoltage surge protectors are installed. The surge arresters protect the GIL from lightning strikes into the overhead line. This protection system has protected the GIL from overvoltages for over 35 years of operation [150]. Today, after 36 years of operation, the GIL has no ageing effects.

7.2.2   Windhoek, Namibia, 1977

The GIL installation at Windhoek, Namibia in 1977 was the first vertical installation in a shaft of a hydropower plant. With this project it was shown that GIL can also be built in long vertical shafts. The conductors are held by conical insulators and the enclosure is fixed to the shaft wall through a steel structure. See the view into the vertical shaft in Figure 7.4.

The total height of the vertical shaft is 115 m. From the high-voltage transformer in the cavern the GIL connects the power to a substation above the cavern, as shown in Figure 7.5.

The technical data is shown in Table 7.3.

In this project the requirement is to have a vertical connection of 115 m which can be solved by GIL. The cavern power plant at Windhoek, Namibia, operates under 245 kV rated voltage and at a rated current of 630 A. The GIL is vertically mounted in the 115 m-deep shaft. The solid design of aluminium pipes and gas as an insulation medium makes an easy assembly possible. This GIL has a total pipe length of 800 m and has been in operation since 1977. See Figure 7.5.

Table 7.3   Technical data, Windhoek, 1977.
Reproduced by permission of © Siemens AG

Rated voltage U_r	245 kV
Rated current I_r	630 A
Rated impulse withstand voltage U_{BIL}	1050 kV
Rated short-time current I_s	40 kA/3 s
Length per phase	0.8 km
Insulating gas	100% SF_6

Figure 7.4 Vertical GIL, Windhoek, 1977. Reproduced by permission of © Siemens AG

7.2.3   Joshua Falls, USA, 1978

This GIL was the first directly buried version world-wide. To minimize the overall visual impact of a new substation, GIL was chosen to connect the GIS with the overhead line at an access point far away from the substation. In Figure 7.6 the open trench is shown during the installation and laying process.

There are two systems with a total GIL length of 1.7 km. The routing required multiple changes of elevation and also horizontal directional changes with angle units, as shown in Figure 7.7.

The radius of the directional change is about 250 m. Besides the passive corrosion protection (the white coating in Fig. 7.6), a cathodic protection system is added to prevent aluminium from corrosion. Since 1978 this has been successfully provided. The electrical potential for

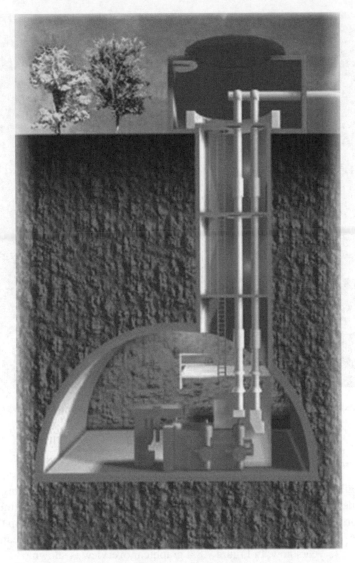

Figure 7.5   Graphical overview of vertical GIL. Reproduced by permission of © Siemens AG

aluminium pipe protection is delivered by polarization cells at each end of the GIL system, isolated to ground to generate the correct potential voltage. The GIL is solidly grounded and cross-bonded at each end to permit induced currents in the enclosure and to minimize the magnetic field.

The GIL sections have 18 m shipping length and are welded on-site for joint connections. Welded joints are always used for directly buried GIL. The technical data is shown in Table 7.4.

Figure 7.6 Joshua Falls, USA, 1978: laying process. Reproduced by permission of © CGIT Systems

7.2.4 Bowmanville, Canada, 1985–7

The GIL installation at Bowmanville in Canada has the highest current rating world-wide. Installed in the years 1985–7, the maximum rated current is 8000 A continuously and a maximum short-circuit current of 100 kA. With different GIL connections from the GIS in the building to the overhead lines, a total of 2.5 km of GIL was installed and has been successfully in operation for 15 years now. In Figure 7.8 a graphical overview is shown.

Table 7.4 Technical data, Joshua Falls, 1978.
Reproduced by permission of © CGIT Systems

Rated voltage U_r	145 kV
Rated current I_r	2000 A
Rated impulse withstand voltage U_{BIL}	650 kV
Rated short-time current I_s	63 kA/3 s
Length per phase	1.7 km
Insulating gas	100% SF_6

Figure 7.7 Joshua Falls, USA, 1978: vertical and horizontal direction changes. Reproduced by permission of © CGIT Systems

The GIL connection from the GIS in the building to the overhead lines is installed on steel structures. The ambient outdoor conditions require low temperatures, ice load and wind load. See the photo of the above-ground installed GIL in Figure 7.9 inside the substation building for the GIS and in Figures 7.10 and 7.11 outside the substation building.

The technical data is given in Table 7.5.

Table 7.5 Technical data, Bowmanville, Canada, 1985–7.
Reproduced by permission of © Siemens AG

Rated voltage U_r	550 kV
Rated current I_r	4000/6300/8000 A*
Rated impulse withstand voltage U_{BIL}	1550 kV
Rated short-time current I_s	100 kA
Length per phase	2.5 km
Insulating gas	100% SF_6

*Depending on substation section.

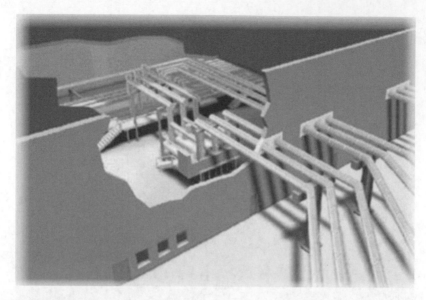

Figure 7.8 Overview of Bowmanville, Canada, 1985–7. Reproduced by permission of © Siemens AG

Figure 7.9 GIL inside the substation building. Reproduced by permission of © Siemens AG

Figure 7.10 GIL outside for connection to the overhead lines. Reproduced by permission of © Siemens AG

7.2.5 Shin-Meika Tokai Line, Japan

In Japan, the longest GIL has been installed in a tunnel over a length of 3.3 km. The operating voltage is 275 kV and the rated current is 6000 A. The GIL has been in operation since 1998 [163–166]. This GIL was assembled on-site by welding 14 m-long sections in the tunnel and the complete transmission line was tested on-site by high-voltage tests [167].

In Table 7.6 the main technical data of the Shin-Meika Tokai Line are given.

The route plan in Figure 7.12 shows the Shin-Meika Tokai Line as a 3.3 km-long connection from Tokai Substation (Tokai S/S) on the Japanese mainland to the Shin-Nagoya Power and

Table 7.6 Technical data, Shin-Meika Tokai Line, Japan. Reproduced by permission of © Chubu Electric Power Co., Inc.

Rated voltage U_r	275 kV
Rated current I_r	6300 A
Rated impulse withstand voltage U_{BIL}	1050 kV
Power frequency test voltage	460 kV
Length per phase	9.9 km
Insulating gas	SF_6

Figure 7.11   Above-ground GIL on steel structures. Reproduced by permission of © Siemens AG

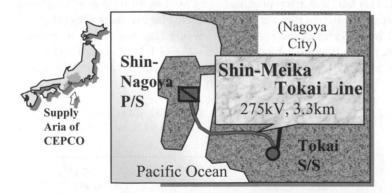

Figure 7.12   Route plan of the Shin-Meika Tokai Line, Japan. Reproduced by permission of © Chubu Electric Power Co., Inc.

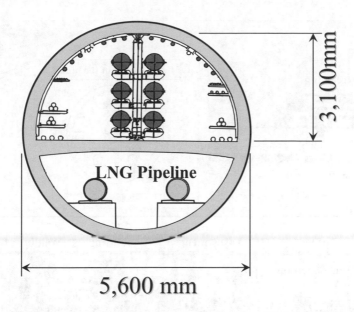

Figure 7.13 Cross-section of double-tunnel system. Top: two GIL systems. Bottom: two liquid gas pipelines. Reproduced by permission of © Chubu Electric Power Co., Inc.

Substation (P/S) at the peninsula of Shin Nagoya in the Nagoya Bay of the Pacific Ocean. From the mainland the Shin-Meika Tokai Line is laid in a double-tunnel system with two liquid natural gas pipelines (LNG) at the bottom of the tunnel and two three-phase GIL in the upper part of the tunnel, as shown in Figure 7.13.

In Figure 7.14 the assembly process of the GIL is shown. Through a shaft on the left-hand side of the graphic, the pre-manufactured GIL sections are delivered to the receiving area and then brought through the shaft down into the tunnel by a crane. In the tunnel a rail transportation system transports the GIL segments to the assembly location along a 3.3 km-long tunnel. There was only one delivery shaft for the GIL sections. In the graphic, the four steps of the assembly process are shown next.

First the adjustments are fixed to the tunnel. In the tunnel centre steel structures are fixed to the floor and roof of the tunnel, which will carry the single-phase insulated GIL pipes. In the jointing, the GIL segments are jointed by placing the conductor in a plug and socket system and by preparing the enclosure pipe for the welding process. The third process step is the welding of the enclosure pipe using an arc welding process. Finally, in the gas charging process step the GIL is filled with insulating gas and is ready for the on-site high-voltage test.

In Figure 7.15 a view into the GIL tunnel section is shown before the GIL was installed. In the middle of the tunnel the steel structures for fixing the GIL sections are shown. Left and right a rail transportation system is shown to transport the GIL sections along the 3.3 km tunnel.

The GIL sections are laid at half-pipe fixing points, which allow axial movement for thermal expansion. In Figure 7.16 a steel structure is shown. On the right side of the photo a clean booth for assembling the conductor plug and socket connection and preparing the outer enclosure for the welding process is shown.

Each GIL section was pre-assembled in a factory and transported by trucks to the delivery shaft on-site. A 14 m-long GIL section is shown in Figure 7.17. Special transportation

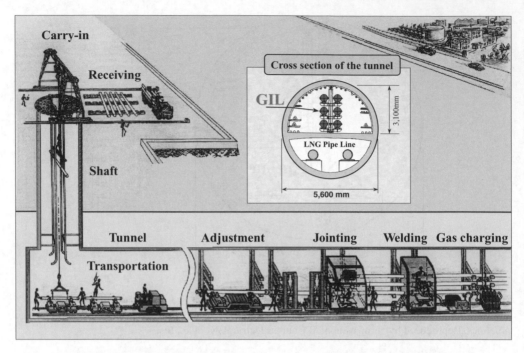

Figure 7.14 Shin-Meika Tokai Line, assembly process. Reproduced by permission of © Chubu Electric Power Co., Inc.

Figure 7.15 View into the GIL tunnel section. Reproduced by permission of © Chubu Electric Power Co., Inc.

Figure 7.16   View of a steel structure to fix the GIL section. Reproduced by permission of © Chubu Electric Power Co., Inc.

Figure 7.17   GIL section ready for laying, 14 m length. Reproduced by permission of © Chubu Electric Power Co., Inc.

tools were developed to avoid any damage to the GIL section, e.g. by shocks. The photo in Figure 7.17 shows the GIL section just before releasing it into the shaft to the tunnel for laying.

The Shin-Meika Tokai Line, at 3.3 km, is the longest GIL installation in the world. The high-current (6300 A) installation connects the gas turbine power plant located at Shin Nagoya with the Tokai Substation to deliver electrical energy into the Nagoya high industrial region. The GIL has operated since 1998 without any interruption or problem. With 2850 MVA in two circuits, this installation makes a large contribution to the electric power supply in the Nagoya region.

Given the requirement for gas pipelines in combination with GIL, a tunnel with two halves and a diameter of 5.6 m was designed. The upper 3.1 m was used for the GIL. A narrow working space and a three-dimensional curve in the horizontal and vertical directions – in dusty surroundings – had to be resolved during the on-site assembly process. The plug and socket joint system of the conductor, a fully automated welding process, the use of a clean room working booth and the modular jointing process of GIL sections provided a very high reliability of the assembly process and failure-free operation.

Now, after more than 12 years of operation, the GIL has proven its very high reliability in transmitting large amounts of electric power.

7.2.6 PALEXPO, Switzerland, 2001

The GIL installation at PALEXPO in Geneva, Switzerland in 2001 was the first N_2/SF_6 gas mixture installation world-wide. The application here is only for transmission as an underground part of an overhead line. This new gas mixture technology was developed for long-distance underground transmission and the PALEXPO project was its first application [168]. Now, almost 10 years later, the project operates reliably without problems. See also Section 2.2.2.

The task was to replace the existing overhead line at the PALEXPO exhibition area at Geneva Airport with an underground connection in a tunnel. Figure 7.18 shows the overhead line to be replaced.

The routing of the GIL is shown in Figure 7.19.

The new exhibition hall (Hall 6) was the reason for the underground transmission between Pylon 175 and 176 over a length of about 470 m. The top view in Figure 7.19 shows that the underground tunnel leaves the overhead line route to the side. The reason for this is that the tunnel is built under a road parallel to a highway, which gave the best access to the site works of the tunnel built from the top using an open trench method.

The opening for erection (1) is the shaft for tunnel access during erection. All GIL units were lowered through this shaft into the tunnel for jointing. The straight unit (2) of the GIL was bent to follow the tunnel curvature. The minimum bending radius is $R \geq 400$ m. Expansion joints (3) take care of the thermal expansion of the GIL, which has free movement at the steel support structures (5). In the middle of the tunnel the GIL is fixed to the fixed support (6). The directional changes are made by angle units (3) of 90° angle and the end of the GIL is gas-tight closed to form one gas compartment using a disconnecting unit (4). The disconnecting units are then connected to SF_6/air bushings on top of the tunnel to connect the GIL to the overhead line at Pylon 175 and 176.

Figure 7.18 PALEXPO, Geneva: replacement of overhead line. Reproduced by permission of © Alpiq Suisse SA

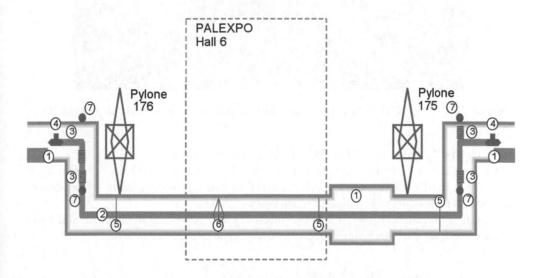

Figure 7.19 PALEXPO, overview from the top. Reproduced by permission of © Alpiq Suisse SA

Table 7.7   Technical data, PALEXPO, Geneva, Switzerland.
Reproduced by permission of © Alpiq Suisse SA

Rated voltage U_r	220 kV
Rated current I_r	2000 A
Rated impulse withstand voltage U_{BIL}	850 kV
Rated short-time current I_s	50 kA
Length per phase	3.8 km
Insulating gas	80% N_2/20% SF_6

Figure 7.20   PALEXPO, view into the tunnel. Reproduced by permission of © Alpiq Suisse SA

The technical data for the PALEXPO project are given in Table 7.7.

In Figure 7.20 a view into the tunnel shows two GIL systems and on the right two angle units of 90°. Elastic bending was used for the first time in this project to follow the bending of the tunnel. The bending radius in this section is about 700 m.

The GIL application of the PALEXPO project was built according to the process of long-distance applications. This laying process has been explained in Section 3.7.

7.2.7   Baxter Wilson Power Plant, USA, 2001

The Baxter Wilson Power Plant was installed in 2001 in Mississippi, USA, to cross a river section which could be flooded at high water levels in the river. The GIL underpasses several 550 kV and 242 kV lines and offered an economical solution. To overpass the 550 kV line would require very high, expensive and (in hurricane regions) also risky solution.

Figure 7.21 Baxter Wilson GIL, USA, 2001: overview under 550 kV and 242 kV OHL. Reproduced by permission of © CGIT Systems

In Figure 7.21 the GIL is shown underneath the 550 kV and 242 kV overhead lines. The GIL was installed on steel structures to be higher than the expected high level of the water, and fenced to protect against vandalism.

The technical data for the Baxter Wilson GIL are given in Table 7.8.

7.2.8 Sai Noi, Thailand, 2002

The Sai Noi substation is an important station for the Bangkok power supply. An extension of an existing GIS required also some long connections between the existing GIS, an existing AIS and the new extended GIS. About 1 km of system length, or 3 km of pipe length, had to

Table 7.8 Technical data, Baxter Wilson GIL, USA, 2001. Reproduced by permission of © CGIT Systems

Rated voltage U_r	550 kV
Rated current I_r	4500 A
Rated impulse withstand voltage U_{BIL}	1550 kV
Rated short-time current I_s	80 kA
Length per phase	3.75 km
Insulating gas	100% SF_6

Figure 7.22 Overview of GIL installation. Reproduced by permission of © EGAT and © Siemens AG

be installed for this reason. This GIL was the first gas mixture GIL for 550 kV with 40% N_2 and 60% SF_6.

The transmission requirements were high. At a rated voltage of 550 kV, a rated current of 4000 A was needed to fulfil the requirements of the transmission lines connected to this substation. At the same time the ambient air temperature was as high as 50°C, and at the same time the direct sun radiation was high too. The GIL can fulfil all these requirements and the operation started successfully in September 2002. An overview of the GIL installation is given in Figure 7.22.

The GIL at Sai Noi was installed according to the same procedures as the PALEXPO project in Geneva, Switzerland. The methods of orbital welding, gas mixture, on-site GIL segment assembly and the pulling method have been applied [74].

Table 7.9 Technical data, Sai Noi, Thailand. Reproduced by permission of © Siemens AG

Rated voltage U_r	550 kV
Rated current I_r	4000 A
Rated impulse withstand voltage U_{BIL}	1550 kV
Rated short-time current I_s	63 kA/1 s
Length per phase	3.5 km
Insulating gas	40% N_2/60% SF_6

Figure 7.23 Assembly tent at Sai Noi. Reproduced by permission of © EGAT and © Siemens AG

Figure 7.24 Inside the assembly tent at Sai Noi. Reproduced by permission of © EGAT and © Siemens AG

The extreme external conditions of very high ambient temperature, extreme sun radiation and high rated current of this important power supply line to Bangkok shows the high power transmission capability of GIL even under extreme climatic conditions.

The on-site assembly was carried out inside an assembly tent. All machinery for the welding preparation, the pre-assembly of straight, angle, compensator and disconnecting unit of the GIL was carried out in the assembly tent. To provide good working conditions the tent was air-conditioned. Internal cleanliness was provided by a closed and controlled access system for personnel and materials.

Figure 7.23 shows an assembly tent at Sai Noi.

In Figure 7.24 a view into the assembly tent is shown, where straight units of the GIL are prepared for the welding process.

The technical data for Sai Noi, Thailand is shown in Table 7.9.

7.2.9   PP9, Saudi Arabia, 2004

The GIL installation at PP9 in Saudi Arabia in 2004 was built to connect 8 blocks of power plants to the substation at 400 kV transmission voltages. The installation is above ground on steel structures and the main reason for choosing GIL is the severe climatic conditions in the desert with high humidity and dust [169].

The high reliability of the GIL and its low transmission losses are the main reasons for GIL. The total length of the single-phase GIL adds up to 17 km. In Figure 7.25 an impressive view of the ~ 5 m above-ground installed GIL is shown.

The technical data for the PP9 project in Saudi Arabia are given in Table 7.10.

Figure 7.25   Ground view of PP9, Saudi Arabia, 2004. Reproduced by permission of © Alstom Grid

Table 7.10 Technical data, PP9, Saudi Arabia, 2004.
Reproduced by permission of © Alstom Grid

Rated voltage U_r	420 kV
Rated current I_r	1200 A at 55°C
Rated impulse withstand voltage U_{BIL}	1425 kV
Rated short-time current I_s	63 kA
Length per phase	17 km
Insulating gas	100% SF_6

7.2.10   Cairo North, Egypt, 2005

The GIL installation at Cairo North in Egypt is a power plant extension, and the connecting GIL is installed at a great height to overpass existing equipment. The commissioning of the 245 kV GIL at Cairo North Substation was in 2004. Extreme environmental conditions of dust, sand and humidity were the main reasons for the GIL in this case. In Figure 7.26 the connection from the power plant to the substation is shown. In Figure 7.27 the connection to the substation building is shown.

The technical data for Cairo North in Egypt are shown in Table 7.11.

Figure 7.26   245 kV GIL Cairo North. Reproduced by permission of © Siemens AG

Table 7.11 Technical data, Cairo North, Egypt.
Reproduced by permission of © Siemens AG

Rated voltage U_r	245 kV
Rated current I_r	3150 A
Rated impulse withstand voltage U_{BIL}	1300 kV
Rated short-time current I_s	50 kA
Length per phase	3.5 km
Insulating gas	100% SF_6

Figure 7.27 245 kV Cairo North Connection to the substation building. Reproduced by permission of © Siemens AG

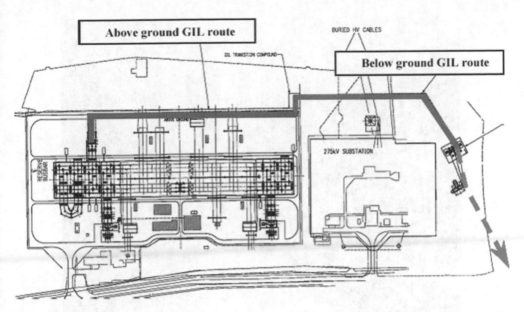

Figure 7.28 Overview Hams Hall above and underground GIL, 2005. Reproduced by permission of © Siemens AG

7.2.11 Hams Hall, Midlands, UK, 2005

The Hams Hall substation is an important link for the London area power supply. To upgrade the substation from 275 kV to 400 kV, new transmission lines needed to be connected. To connect these new lines with the GIS, a GIL was used to underpath existing overhead lines as shown in Figure 7.28.

The green line section on the left in the graphic is installed above ground on steel structures and inside the substation. Outside the substation fence the GIL goes underground into a trench which is covered by a concrete plate. At the fenced overhead line tower the GIL leaves the trench and is connected by bushing installed in the overhead line tower to the overhead line.

The technical data is shown in Table 7.12.

Table 7.12 Technical data, Hams Hall, Midlands, UK, 2005. Reproduced by permission of © Alstom Grid

Rated voltage U_r	420 kV
Rated current I_r	3150 A
Rated impulse withstand voltage U_{BIL}	1425 kV
Rated short-time current I_s	63 kA
Length per phase	1.5 km
Insulating gas	80% N_2/20% SF_6

Figure 7.29 Huanghe Laxiwa, China, 2009 Horizontal tunnel. Reproduced by permission of © CGIT Systems

7.2.12 Huanghe Laxiwa, China, 2009

The high power transmission requirement at the Huanghe Laxiwa hydropower station in China is the first 800 kV GIL which has been installed inside a tunnel system of vertical and horizontal sections. The vertical section is 209 m and the total single phase length of the GIL is 2.75 km. In Figure 7.29 a view into the horizontal tunnel section with two 800 kV GIL systems installed at the tunnel walls is shown.

In Figure 7.30 a view into the vertical tunnel section with two 800 kV GIL of 209 m height is shown. The GIL single phases are fixed by steel structures at the tunnel walls.

The technical data for the Huanghe Laxiwa hydropower plant in China are shown in Table 7.13.

Table 7.13 Technical data, Huanghe Laxiwa, China, 2009. Reproduced by permission of © CGIT Systems

Rated voltage U_r	800 kV
Rated current I_r	4000 A
Rated impulse withstand voltage U_{BIL}	2100 kV
Rated short-time current I_s	63 kA
Length per phase	2.75 km
Insulating gas	100% SF_6

Figure 7.30   Vertical tunnel, Huanghe Laxiwa, China, 2009. Reproduced by permission of © CGIT Systems

7.2.13   Kelsterbach, Germany, 2010

The extension of the Frankfurt International Airport required undergrounding the existing overhead lines of 220 kV and upgrading the transmission voltage to 400 kV. In Figure 7.31 a construction site view shows the overhead lines and the directly buried GIL. See also Section 2.2.2.4. In the middle of the photo the on-site assembly tent is shown, which is placed on top of the laying trench. The GIL sections are assembled and welded in the tent and then pulled into the trench [170].

A view inside the assembly tent is given in Figure 7.32. The straight units and angle units are pre-assembled before laying inside the tent.

The straight units are welded on-site using the automated orbital welding process. See Figure 7.33.

The straight units were welded in the assembly tent and then pulled into the trench. In Figure 7.34 a trench with six single-phase GIL pipes for two three-phase systems is shown before the trench was closed.

Figure 7.31 Overview of directly buried GIL at Kelsterbach, Germany, 2010. Reproduced by permission of © Siemens AG

Figure 7.32 Inside the assembly tent. Reproduced by permission of © Siemens AG

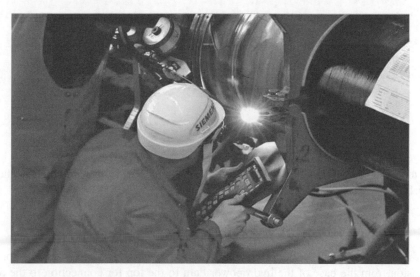

Figure 7.33 Kelsterbach, Germany on-site welding. Reproduced by permission of © Siemens AG

Figure 7.34 Kelsterbach, Germany six single-phase GIL pipes in the trench before closing. Reproduced by permission of © Siemens AG

Table 7.14 Technical data, directly buried GIL, Kelsterbach, Germany, 2010.
Reproduced by permission of © Siemens AG and © Amprion Gmbh

Rated voltage U_r	420 kV
Rated current I_r	3000 A
Rated impulse withstand voltage U_{BIL}	1425 kV
Rated short-time current I_s	63 kA
Length per phase	6 km
Insulating gas	80% N_2/20% SF_6

The technical data for the directly buried GIL using N_2/SF_6 gas mixture is shown in Table 7.14.

7.2.14 Xiluodu, China, 2011

The hydropower station at Xiluodu in China requires seven vertical GIL systems for power transmission from the base of the hydropower dam to the top for connection to the 550 kV network. The hydrodam is shown in the graphic in Figure 7.35.

In Figure 7.36 the principle of the vertical GIL is shown. The total height of 620 m is covered in two sections, vertically inside the dam. The total length of the single-phase GIL of the seven three-phase systems adds up to a length of 12 km.

In Table 7.15 the technical data for the vertical GIL at Xiluodu is shown.

Figure 7.35 Overview of hydropower station, Xiluodu, China, 2011. Reproduced by permission of © Siemens AG

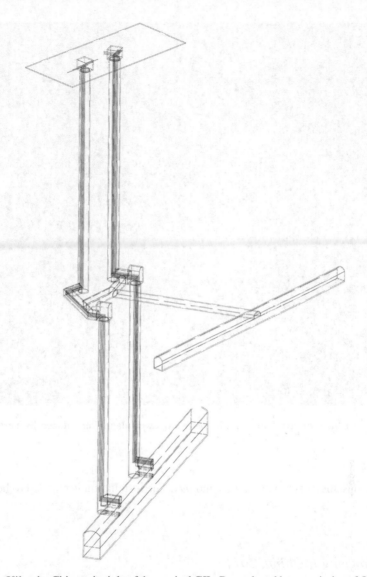

Figure 7.36 Xiluodo, China principle of the vertical GIL. Reproduced by permission of © Siemens AG

Table 7.15 Technical data, Xiluodu, China, 2011.
Reproduced by permission of © Siemens AG

Rated voltage U_r	550 kV
Rated current I_r	4500 A
Rated impulse withstand voltage U_{BIL}	1550 kV
Rated short-time current I_s	63 kA
Length per phase	12 km
Insulating gas	100% SF_6

Figure 7.37   Overview of river, Jingping I, for hydropower plant. Reproduced by permission of ©
Siemens AG

The GIL is installed using welding technology. The installation is planned to be completed
in 2013.

7.2.15   Jingping I, China, 2011

Another hydropower station at Jingping I in China requires vertical GIL inside a dam. The dam
is built on a river in China, in a remote area to bring regenerative power to the metropolitan
areas. In Figure 7.37 an overview of the river is given.

The principle of the Jingping I GIL installation in the dam of the hydropower station is shown
in Figure 7.38. The GIL connects the generation in the dam with the 550 kV transmission
network outside the dam through vertical shafts in the tunnel. The total single-phase length
of the GIL is 3.2 km. The GIL is SF_6 insulated and the GIL sections are jointed by welding
technology.

The technical data for the Jingping I GIL project are shown in Table 7.16.

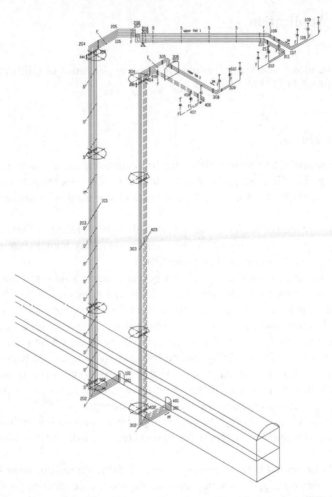

Figure 7.38 Principle of the GIL inside the dam. Reproduced by permission of © Siemens AG

Table 7.16 Technical data, Jingping I, China, 2011.
Reproduced by permission of © Siemens AG

Rated voltage U_r	550 kV
Rated current I_r	4000 A
Rated impulse withstand voltage U_{BIL}	1550 kV
Rated short-time current I_s	63 kA
Length per phase	3.2 km
Insulating gas	100% SF_6

7.3   Future Application

7.3.1   General

In this section we look to the future for possible new applications of GIL in the high-power transmission network [173, 174, 203].

7.3.2   Traffic Tunnels

The GIL can be added in a street or railroad tunnel, because it causes no damage to the other users of the tunnel. The GIL can be fixed on the tunnel roof on steel structures with very long fix point distances (30–40 m), or laid in a separate tunnel also used as an emergency exit tunnel [174–177, 210].

The low capacitive load of the GIL makes it possible to build lengths of up to 80 km without any compensation of the phase angle. For this reason, very long tunnel systems for trains, as planned in the Alps, can also be equipped with GIL.

The gas-insulated transmission line is a high-power transmission system using aluminium pipes for the conductor and the enclosure. The insulation medium is an N_2 and SF_6 gas mixture. The solid, metallic enclosure has a strong mechanical strength against outer and inner forces, and withstands even the extreme situation of an internal arc. Only very little material distortion can be seen inside the enclosure, even after an arc current of 63 kA and a duration of 0.5 s. The reason for this arc withstandability can be found in the way an arc burns in N_2/SF_6 gas mixtures. The nitrogen causes, similarly to in air, a very wide footprint of the arc and therefore low temperatures on the enclosure surface. In consequence, almost no melting of material occurs. The huge gas compartments keep the pressure increase low.

This physical behaviour of the N_2/SF_6 gas-insulated transmission lines allows new applications in combination with publicly accessible tunnels (e.g., railroad or street tunnels as shown in Figure 7.39).

GIL can safely be routed through tunnels carrying traffic on rails or streets. The GIL has a solid metallic enclosure and does not burn or explode, as explained in Chapter 3. The combinations of GIL and street or railroad tunnels are shown in Figure 7.39. Three examples are given. The first one is a traffic tunnel with cars and a GIL mounted on top of the tunnel; the second is a double railroad tunnel system with a separate GIL tunnel; the third example is a double-track railroad tunnel with a GIL included.

The use of such traffic tunnels with GIL is now under investigation in different parts of the world. In the European Alps, interconnections between Germany, Austria, Switzerland, Italy and France are now planned to improve the traffic flow and to allow trade of electrical energy. In China and Indonesia, interconnections between the mainland and islands or between outlying islands are under investigation. In the near future, GIL will become economically viable and will be widely used as high-power, long-distance transmission lines [5, 9, 10].

7.3.3   Roads and Highways

For long-distance applications the GIL can be combined with roads and highways. Figure 7.40 shows the GIL laid under a two-lane road with a total width of 5 m. The distance between the

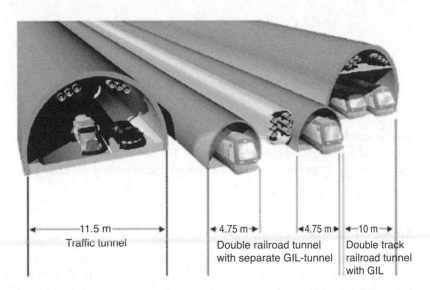

←————11.5 m————→   ◄ 4.75 m ►   ◄4.75 m►   ◄—10 m—►
Traffic tunnel   Double railroad tunnel   Double track
                                with separate GIL-tunnel railroad tunnel
                                                          with GIL

Figure 7.39   GIL in typical traffic tunnels. Reproduced by permission of © Siemens AG

pipes is 1.5 m and the minimum soil coverage above the pipes is 1.0 m [173], see Table 7.17 for dimensions data.

The two GIL systems will allow twice 2000 MW at 400 kV rated voltage. Because of the low magnetic fields outside the GIL, the impact on electronic systems in cars is low. The low capacitive load of the GIL allows long lengths without the need for phase angle compensation. Typical lengths of 80 km without compensation are possible, depending on the connected

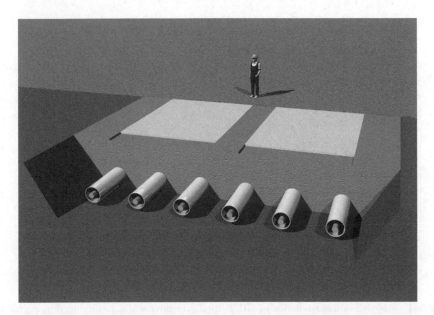

Figure 7.40   GIL directly buried under a two-lane road. Reproduced by permission of © Siemens AG

Table 7.17 Dimensions of the GIL under the road

Trench width	5 m
Distance between pipes (centre line)	1.5 m
Minimum soil coverage	1.0 m

network conditions. The high safety level even in case of internal failure allows the GIL application in combination with public traffic [4, 6]. The low transmission losses of the GIL limit the maximum soil temperature below the limiting temperature of 40°C [76, 79, 147, 178, 179].

7.3.4 Above-Ground and Cross-Country

In this section some ideas are presented for laying options in applications outside substations. When leaving the fenced area of a substation, the GIL needs some protection from vandalism and dangerous manipulations of this high-voltage, high-power transmission system [180]. One possibility is shown in Figure 7.41. The GIL is protected by lightweight walls from public access. The walls are used to fix the GIL.

7.4 Case Studies

The future needs of high-power interconnections are a challenge to the transmission system. GIL offers some interesting solutions.

Figure 7.41 Above-ground installation of GIL behind protective walls. Reproduced by permission of © Siemens AG

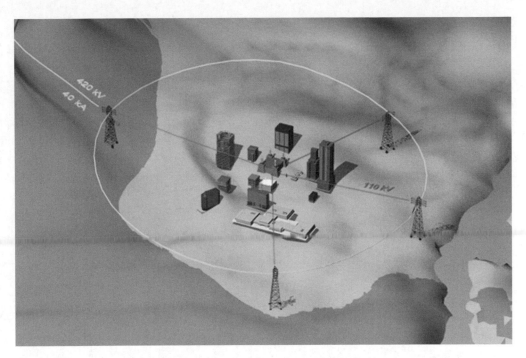

Figure 7.42   Power supply to metropolitan areas in 1970. Reproduced by permission of © Siemens AG

7.4.1   Case Study: Metropolitan Areas

Metropolitan areas world-wide are growing in load density, mainly at their centres. Demand for power has grown because of the construction of huge residential and tall office buildings with air-conditioning and lots of electronic equipment, leading to increases in electric loads of up to 10% per year in metropolitan areas [173]. The following short historical overview explains how the power supply to metropolitan areas has developed over the last 30 years. Figure 7.42 shows the principles of power supply in a metropolitan area. Power generated in a rural area is connected to a metropolitan area by 400 or 500 kV overhead lines with a short-circuit rating of 40 kA. Several substations are placed around the city as overhead towers using a bypass or a ring structure around the metropolitan area, from which 110 kV cables transport electrical energy into the centre of the city, where medium-voltage energy is distributed [81, 151].

In Figure 7.43 the metropolitan area has grown, with more tall buildings in the centre. Most cities still have a 400/500 kV ring around the city, but the short-circuit rating has been increased to 50 kA or, in some places, to 63 kA. Note the second connection to the ring from another rural power generation area. More 110 kV cables are connected to the ring to transport the energy to the substations in the city for distribution. Moreover, world-wide experience with very high short-circuit ratings shows that short-circuit rating values cannot go far above 63 kA because of mechanical problems. So, the only way to increase the power transportation into the city is to lay 400 kV underground bulk power transmission systems right into the centre. In such cases, the GIL offers a good solution.

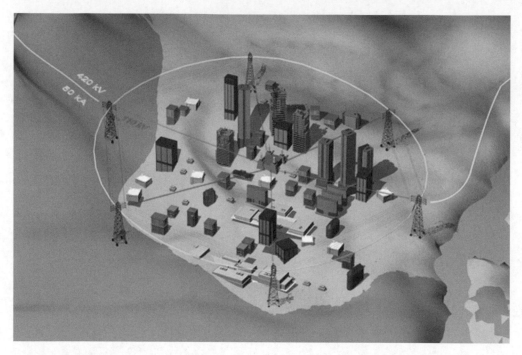

Figure 7.43   Power supply to metropolitan areas in 2000. Reproduced by permission of © Siemens AG

Figure 7.44 shows the metropolitan area as it may appear in 2020. The same metropolitan area with more buildings has grown further, and a 400/500 kV, 63 kA, double-system GIL is built as a diagonal connection underpassing the total metropolitan area. This GIL could have a length of 30 to 60 km. The solution illustrated here allows splitting of the short-circuit ratings of the ring into two half rings, connected directly to the 400 kV high transmission power GIL in the centre of the metropolitan area.

A further big advantage for such new applications is the possibility of sharing the investment of the tunnel between the electric power transmission company and the traffic company. The load density in the centres of metropolitan areas has been growing by 7–10% per year over recent years. The reason for this is the erection of new tall buildings for offices or living, even complete city complexes including shopping and theatres have been built world-wide in city centres. The existing power supply to metropolitan areas world-wide is typically based on a 420/550 kV ring or bypass with substations connecting the city centre with 123/275 kV cable connections, as shown in Figure 7.44.

Today, the 123/275 kV voltage level in many cities is at the end of the transmission capability and the space for new tunnels or trenches is not available. One solution is to have a GIL with a strong high-power transmission line diagonally across the metropolitan area, with rated voltages of 400/500 kV and rated currents of 3150/4000 A. With GIS, connections to this high-power line can be made in the centre of the city to serve the customers.

In a short historical overview, we now explain how the power supply to metropolitan areas has developed over the last 30 years.

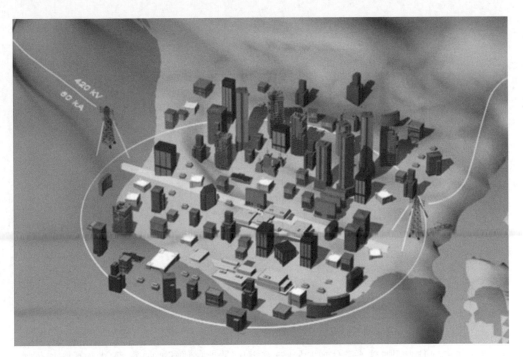

Figure 7.44 Power supply to metropolitan areas in 2020. Reproduced by permission of © Siemens AG

7.4.2   Case Study: London

As an example for GIL applications, which are under consideration now, we use the city of London (see Figure 7.45) [81, 181].

London has a 400 kV overhead line ring around the city with a diameter of about 80 km. Along that ring several power generation plants are situated, and also 400 kV transmission lines are connected to deliver energy into the overhead ring line for the London metropolitan area. Starting from that overhead ring line, some spur lines transport the electrical energy to an inner 275 kV cable ring. Starting from the cable ring, the distribution cables of 132 kV and below are connected to bring the energy to the end user in the city centre. Today, the spur lines from the 400 kV to the 275 kV ring are very heavily loaded. New overhead lines cannot be built because of public non-acceptance. The GIL could offer a long-term solution for economic and reliable power supply to the London city centre by offering new underground, high-power and long-length transmission lines. Using existing rights of way it would be possible to build new GIL lines underneath the existing overhead lines in a directly buried laying technique with a power transmission rating of 2800 MVA per system. Also, new tunnels of 3 m diameter equipped with two systems of GIL could bring 4400 MVA of electric power into the city centre, 220 MVA with each GIL system. This could solve most of the power delivery needs of the city centre. London is only one example of a big metropolitan area, the same electric energy delivery problems can be shown in any big metropolitan area.

To solve the problem of future increases in electrical loads in city centres, a GIL connection crossing the whole centre could be a solution. The low magnetic field outside the GIL is

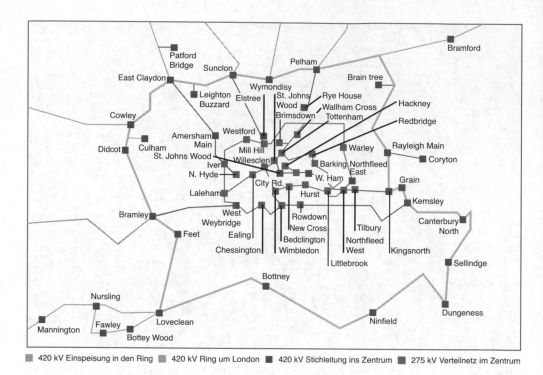

Figure 7.45 Power supplies in the London conurbation: 420 kV outer ring with overhead lines and 275 kV inner ring, usually cables. Reproduced by permission of © Siemens AG

an important advantage in the centre of cities, even if such low values as 1 µT (now law in Switzerland) must be met at street level. The low capacitive load would also allow system lengths of 80 km to be built without electrical compensation of the phase angle.

A strong power connection with rated currents of, for example, 3150 A or 4000 A per system, with very low transmission losses and low magnetic field strength in this densely populated area is a solution for such typical power supply problems around the world. The GIL is also the most economic solution under these conditions [151, 174].

7.4.3 Case Study: Berlin Diagonal

Typical for electric power supply of large cities is that the energy is transmitted to the city centre from large power plants outside the city. Sometimes some hundred kilometres away, using the 400 kV or 500 kV transmission network usually with overhead lines to bring the electric power close to the city or load centre and then distribute the energy by cable of 110 kV to 145 kV voltage levels. Because of the increasing power demand, new distribution lines are needed. Space in the streets is limited to add new cables from the overhead line outside the city to the centre [182–186].

The streets are full of power cables, telecommunication lines, gas pipes, waste water and drinking water. A long-term solution can be offered by a GIL in a tunnel, deep under the

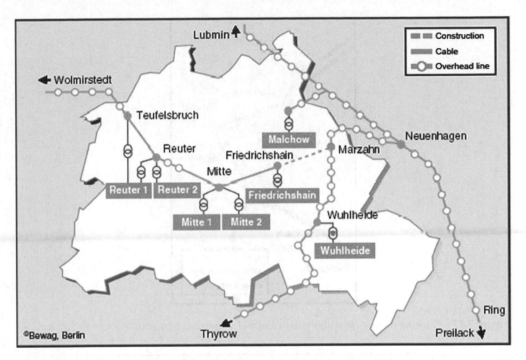

Figure 7.46   Electric power supply in Berlin with 420 kV diagonal. Reproduced by permission of ©
Siemens AG

surface. In Berlin a tunnel of 3 m diameter at 20–40 m depth under the street was built to
create a diagonal power link under the city (see Figure 7.46). Over a length of about 30 km in
sections, a 400 kV bulk voltage connection was created to bring electric power on a 400 kV
voltage level right into the city centre. This concept can be seen as a solution for power supply
to large metropolitan areas.

The principle of diagonal, 400 kV or 500 kV, high-voltage ratings in electric tunnels deep
under the city offers new opportunities for much larger cities like Shanghai, New York or
Mexico City to solve the increasing power needs.

The areas of such large cities reach diagonal lengths of 30–60 km and power ratings of
2000 MW per electric system. The GIL offers such high ratings in tunnels, at low temperatures,
low magnetic field and with high safety levels. Reliable electric power supply of the future
can be obtained with GIL [151, 154].

7.4.4   Case Study: Mountains

To cross mountains, GIL can be installed in tunnels. A case study in Austria shows the
possibility of connecting an existing overhead line on the north side of the Alps through a
tunnel at the south side of the Alps, see Figure 7.47. This allows strong interconnections
between transmission network sections.

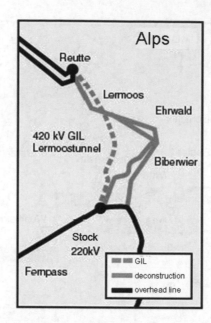

Figure 7.47 Crossing of the Alps with a GIL in a tunnel. Reproduced by permission of © Siemens AG

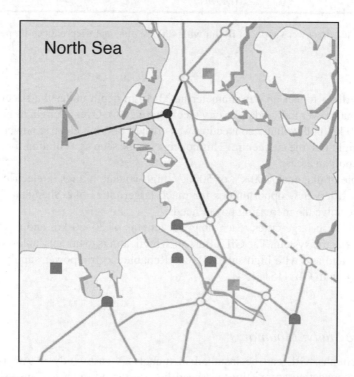

Figure 7.48 Offshore wind park connection to 400 kV transmission systems. Reproduced by permission of © Siemens AG

7.4.5   Case Study: Sea

New 400 kV, high-power transmission lines are needed to connect the offshore wind parks with the transmission net. The North German coastal landscape is very flat with its own beauty. New overhead lines are not acceptable to the public. Additional transmission capability is needed to transport the offshore wind energy plant in the North Sea to the load centres in Southern Germany.

Planning for several Gigawatts of electric power offshore is underway. In the North Sea region only one 400 kV overhead line serves as energy transportation North/South, which is well loaded today. A parallel overhead line would be necessary, but will have problems in being permitted by the public authorities, at least in populated areas. Also, the connection of the offshore wind parks to the continent with a current rating of 4000 A can be solved by GIL. Also in this case, the GIL allows an economical underground solution. See Figure 7.48.

The future environmental requirements to transmission systems are asking for underground solutions, with high power ratings, over long distances, with low magnetic fields, high safety and high reliability. The GIL is the only technical system which combines these requirements [154].

The connections to the offshore wind farm are made by underwater tunnels to collect the wind energy and bring it (concentrated) to the land for connection to the transmission network.

7.4.6   GIL/Overhead Line Mixed Application

A very likely form of application of GIL is in combination with overhead lines. The cost difference between overhead lines and buried GIL will ask for solutions where overhead lines are build where possible and GIL at those location where a permission can not be reached.

This may be close to public living areas or at environmental protected landscapes.

One section or more can be built underground by GIL without limiting the total transmission power of the line. One system overhead line into one underground system GIL.

The connection from the the GIL to the overhead line is made by bushings.

This combination offers new solution possibilities when planning new transmission lines and leads to an optimized solution [211, 212].

8

Comparison of Transmission Systems

8.1 General

The comparison of transmission systems is a complex task. The technical and economical aspects vary greatly from project to project. Many parameters influence the technical layout of the transmission system and the cost. The requirements for a transmission line include the purely technical parameters such as voltage, current, short-circuit rating, overvoltages, temperature limits and electromagnetic field limitations. Also, site condition parameters such as accessibility, maximum transport weight or size, type of soil, transport roads and space for workshops have a major impact on feasibility and cost. Later (in operation) the "soft" parameters such as aesthetics, non-visibility, noises, etc. may be the hardest factors to get permission for operation. Environmental parameters of landscape protection, natural reserve protection and impact on flora and fauna need to be taken into account as they might stop a transmission line project [79].

Besides GIL, mainly overhead lines and solid insulated cables are available today. When only economics are taken into account, then the overhead line offers the solution with the lowest price. But in some cases the overhead line might not be buildable because of public opposition or the non-availability of space. Then the GIL or solid insulated cables can offer an alternative solution. This is often the case in metropolitan areas, city centres, airports or ecologically protected areas.

The GIL offers advantages when high power ratings are required, and one GIL system can be used instead of two cable systems [154].

8.2 GIL Features

The need for high-power supply lines to solve the power demand of metropolitan areas in the 21st century has brought this proven technology more into focus. The requirement came about to use GIL not only for short links but also for power transmission over longer distances. GIL will be used as an alternative to overhead lines where those are not possible or not permitted by public and authority constraints.

Gas-Insulated Transmission Lines (GIL), First Edition. Hermann Koch.
© 2012 John Wiley & Sons, Ltd. Published 2012 by John Wiley & Sons, Ltd.

Table 8.1 GIL features

Safety	• non-flammable gas insulation
	• solid grounded enclosure
Transmission capability	• same power capacity as overhead lines
	• high overload capability
	• low capacitive load
Losses	• low transmission losses
	• negligible insulation losses (tan δ)
	• low phase angle compensation requirement
EMF	• very low electromagnetic field
Network integration	• transmission behaviour like overhead lines
	• direct connection to GIS
	• easy connection to AIS through bushings

GIL is a transmission system which can transmit the full power transmission rating of an overhead line (e.g., 2000 MVA) underground without parallel installations of additional systems and without forced cooling. GIL can be laid directly buried, in a tunnel or above ground. No reactive power compensation is needed for GIL, even for longer transmission distances of 60–80 km. GIL is a safe system. All materials are inflammable and the insulating gas is not toxic. The GIL technology shows almost 40 years of experience of high-voltage applications with very high reliability [79, 154]. In Table 8.1 the features of GIL are listed [10].

8.3 Technical Comparison

8.3.1 General

Many different parameters have to be evaluated to compare technical systems as transmission lines. Each technology (overhead line, cable systems and GIL) has its own features. In comparison, the dependencies between the features make it difficult to find a simple answer as to what would be the best solution. Here only the principles of comparisons can be explained, and quantification can be given if the impact is large or small. Each project has its own set of requirements and the impact of technical features is different. That is why it is necessary to evaluate every project separately.

8.3.2 Losses

The losses of a transmission system can be split into load-related losses, dielectric losses and losses related to operational requirements (e.g., losses of compensation coils or capacitor banks). To show the impact of losses related to power transmission systems, some examples are given [216].

GIL/Cable Losses are directly related to the resistance of the transmission systems. Usually for high power transmission, overhead lines and GIL use aluminium and cables use copper. The cross-section and the specific materials are the basis for the losses, which are shown in Table 8.2 [180].

Table 8.2 Transmission losses at 1100 MVA – an example

Losses	Oil-filled cable (cross-bonding)	XLPE cable (cross-bonding)	GIL (solid grounded)
Cross-section (mm^2)	2000	1600	conductor: 5300 enclosure: 16000
Conductor material	copper	copper	aluminium
Load-independent (dielectric) losses (W/m)	approx. 40	5–10	≈ 0
Load-dependent losses (conductor and shield or enclosure losses) (W/m)	approx. 110	approx. 125	approx. 73
EMF levels	high	high	low

The data for the load-dependent losses apply to a required rated transmission capacity of 1100 MVA. The shield losses of the cables have been optimized by means of cross-bonding, see Table 8.2.

The dielectric, load-independent losses depend essentially on the insulant medium used. They are negligible in the case of GIL when gas insulation is used and the total volume of the solid insulation of the cast resin insulator is small. The load-dependent losses are approximately proportional to the square of the load current and depend on the conductor cross-section and diameter. On account of the large cross-section, the GIL possesses significantly lower conductor losses than the oil-filled or XLPE cable systems. Figure 8.1 shows the transmission losses as a function of the transmission capacity. The consequence of using cross-bonding is that the sheet currents are reduced and therefore the outside magnetic field is higher compared with solidly grounded systems.

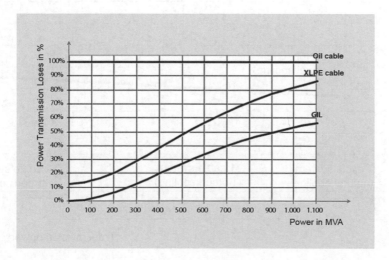

Figure 8.1 Comparison of transmission losses of the 400 kV transmission systems in relation to oil-filled cables. Reproduced by permission of © Vattenfall Europe Netzservice Gmbh

Table 8.3 Physical parameters of the oil-filled, or XLPE cables and GIL [38]

	Oil-filled cable (cross-bonding)	XLPE cable (cross-bonding)	GIL (solid grounded)
Inductance per length L'_b (mH/km)	0.68	0.73	0.22
Capacitance per length C'_b (nF/km)	269	183	54
Resistance per length R' (mΩ/km)	23	19	9.4
Surge impedance Z (Ω), where $Z^2 = L'_b/C'_b$	50	64	63

Table 8.3 shows guide values for the magnitude of the inductance and capacitance per length. It can clearly be seen that the load capacity, which is proportional to the capacitance per length C'_b, in the case of GIL is significantly lower than in the case of cables. This means that, according to system configuration, the expense for current compensation coils can be reduced when using GIL.

The natural transmission capacity is proportional to the characteristic impedance. Here, it can be seen that the GIL evidences similar values to an XLPE cable and thus also possesses a comparable natural transmission capacity.

If only the pure system cost (less operating cost) is considered, it can be seen that for low transmission capacity rates, a naturally cooled cable system evidences lower system cost than a comparable GIL system. A higher transmission capacity can be implemented in a cable system only by means of forced-air cooling, water cooling or a second parallel system. This will also increase the specific system cost. Figure 8.2 illustrates this relationship in a schematic diagram. A naturally cooled GIL system possesses, with higher transmission capacity, lower

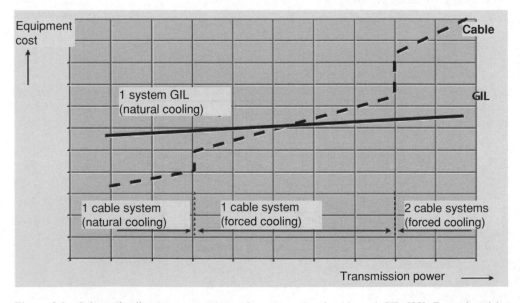

Figure 8.2 Schematic diagram: comparison of system cost of cable and GIL [38]. Reproduced by permission of © Vattenfall Europe Netzservice Gmbh

system cost than a comparable cable system. The comparison is based on a tunnel-laid GIL or a GIL directly buried in normal landscape.

If it is a matter of replacing an overhead power line connection with high transmission capacity of underground construction, the GIL (by virtue of its operating properties and system cost) is an alternative to the cable system [38].

For transmission losses see also Section 4.3.

8.3.3   Magnetic Fields

Magnetic fields are an evaluation criterion if the transmission line is built to permanent living areas or office buildings. The magnetic field is directly related to the current in the conductor and, for solid-grounded GIL or cross-bonded cables, independent of the enclosure or shield current. See Section 5.3.

The overhead line under investigation here has the so-called "Donau tower profile" which holds two 400 kV systems. One on each side in an triangle of two levels. The GIL and the cable are single phase and laid in parallel in the ground with a minimum of 1 m of soil cover. In Figure 5.4 the magnetic field around the transmission system is shown. It can be seen that for GIL, the magnetic field around the system is 10 to 20 times lower [159].

High values of the magnetic field are found with overhead lines around the three phase conductors located in an triangle, as shown in Figure 8.3. The magnetic field value is reduced to the side of the overhead line and reaches values below 5 μT at 40 m distance, see also Figure 5.4.

The highest value of the magnetic field is found with cross-bonded cables, as shown in Figure 8.3. The field value is reduced quickly to the side of the cable and reaches almost zero at 20 m. The peak value right above the cable is close to 30 μT.

The overhead line shows a smaller peak value at about 24 μT but because of the building height of the overhead towers, the magnetic fields are more spread to the sides so that the 5 μT

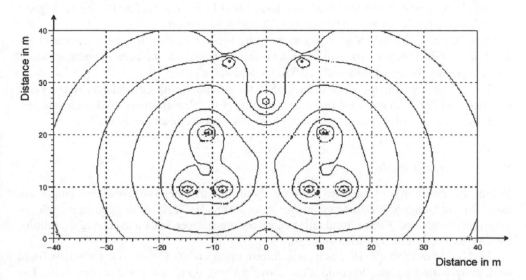

Figure 8.3   Magnetic fields. Reproduced by permission of © Siemens AG

Table 8.4 Maximum voltage rating for AC

Type	AC voltages
Overhead line	1200 kV in India in planning (expected in 2012–2013)
	1100 kV in China in operation
	800 kV in USA, South America, Asia in operation
	550 kV and 420 kV standard value world-wide in operation
XLPE cables	550 kV in operation, e.g. Japan
	420 kV in operation in Europe
GIL	1200 kV in USA test operation in 1980s
	800 kV in China in operation
	550 kV in Thailand in operation
	420 kV in Europe in operation

mark is at 40 m. The lowest value of magnetic fields is with the GIL, where only 2 µT are shown right on top of the GIL.

Magnetic field calculations need to be done for each project as the values are related to the current and the laying option. Here only the principle can be shown.

8.3.4   Voltage Rating

The maximum voltage ratings as installed today are shown in Table 8.4.

The overhead lines in operation use towers of 80–120 m height with 1100 kV or 1200 kV. The size increase of an overhead line tower from 550 kV to 1100 kV is doubling the tower height from about 40 m to 80 m. The size increase for a GIL is from 500 mm diameter to about 650 mm, less than 20%. This means that in cost the higher the voltage ratings the closer are the absolute cost of GIL and overhead lines. The XLPE cables at 550 kV are the highest values used today. Many applications are at 420 kV voltage level.

The GIL had one 1200 kV pilot project in operation in the USA. It is not used any more. The 800 kV GIL is newly built in hydrodam projects in China and has been in operation since 2008. Standard voltages world-wide are 550 kV and 420 kV.

The GIL uses a gas mixture of nitrogen and SF_6 for insulation purposes under an overpressure of 6 bars, similar to an overhead line where the gas of the atmosphere is used for insulation. Therefore, the electrical system behaviour of the GIL and overhead line is similar.

8.3.5   Current Rating

Typical current ratings of overhead lines are about 3000 A when four wires in a bundle are used. Typical ratings of 1600 mm^2 cables are about 2000 A, and 2500 A for 2500 mm^2 cables of the copper conductor. These typical ratings are very much dependent on the laying conditions, e.g. in a tunnel or directly buried in the soil.

The 420 kV and 550 kV GIL has typical current ratings of 3000 to 4000 A for a tunnel-laid or directly buried layout. When installed above ground, up to 8000 A has been built. The current rating is part of the GIL design for each project and its specific requirements [5].

For higher voltage ratings the diameter of the enclosure and conductor is increasing. An 800 kV GIL has about 630 mm enclosure diameter and the current rating is typically 5000 A. The enclosure diameter of a 1200 kV GIL is about 800 mm and the current rating is typically 6000 A.

8.3.6 Short-Circuit Rating

The short-circuit rating is a more complex design criterion because it will affect conductor and ground wires. The design needs to cover electromagnetic forces and thermal heat-up. The electromagnetic forces are between the conductors and the thermal heat-up is in the conductor and the enclosure of the GIL or the shield of the cable or the ground wire of the overhead line. Short-circuit ratings of 63 kA are standard. In some cases 80 kA is reached and even 100 kA in some cases. The solidly grounded GIL and the large cross-section of the enclosure allow such high values for GIL.

The overhead line in the transmission network uses its own ground wire to have a safe ground connection to the next substation for the protection system to trip the line. The limitations of the short-circuit rating are given by the limited conductor and ground wire cross-sections and possible overheating of the wires. The increase in short-circuit rating increases the cost of overhead lines. The cable needs to have cable fault protection to protect the cable from short-circuit ratings because the short-circuit rating is limited. The cross-section of the cable shield is low and may overheat if short-circuit ratings are too high.

8.3.7 Overvoltages

Overvoltages come from the switching operations in the system or by lightning strikes to the overhead line. These overvoltages will enter the overhead line, GIL or cable when connected. The overhead line is designed for overvoltage discharges over the insulators on the towers. GIL and cable are designed to withstand these overvoltages by increased insulation levels. Usually, the GIL and cables which are connected to overhead lines are protected by surge arresters at the transition points. Today's overvoltage surge protection of GIL and cables is very effective and also provides protection for high-frequency transient overvoltage.

8.3.8 Temperature Limits

The temperature limits for overhead lines are given by the insulators with values of 100°C to 120°C depending on the material. Temperature limits are given by the increased length of the wire and limitations of the sag of the wire to keep the distance to ground.

The cable temperature limits are given by the insulation material and the soil maximum temperature or the maximum temperature of the tunnel. The temperature limits are given by the cable along the line or by the joints. Typical maximum temperatures are 40°C to 50°C directly buried or 60°C to 70°C in a tunnel.

The GIL maximum temperatures are given by the insulator material, which is 105°C to 120°C. For directly buried GIL the soil temperature limits are 40°C to 50°C when laid in a tunnel or 60°C to 70°C above ground, according to touch temperatures regulated in standards.

8.4   Site Comparison

The routing of the transmission line has a strong impact on the technical feasibility and the economics. The route planning is a complex investigative and project-related work. Some of the parameters are explained blow.

8.4.1   Accessibility

The accessibility of the route and the site workshop defines the speed of erection or laying. Overhead lines have good site accessibility. The materials for the towers are relatively light and easy to transport. In some cases prefabricated towers are transported by helicopters if no road is available. This is usually not done with cables, which are transported on road and use heavy-weight trucks for transportation. Therefore, site accessibility is sometimes limited by the maximum transport weight of the cable coil.

The GIL is transported in pipes for the enclosure and conductor. The weight of these aluminium pipes is relatively low, but the transport length defines the site accessibility. GIL pipe sections are 12–15 m long to meet the transport requirements on-site.

8.4.2   Maximum Weight

Transmission lines are often built in rural areas with limitations on roads and bridges. Large roads may have a high weight allowance of 40 t but this may go down to 10 t or even 7.5 t on small streets. For cross-country transmission lines these weight limitations are part of the project planning or the route to be chosen.

By basic structure the overhead line has the lowest weight restrictions. The light-weight tower and also the aluminium wires will meet most road requirements in weight.

The GIL aluminium pipes are relatively low weight. With a typical weight of about 40 kg per metre, a 10 m-long GIL has 400 kg. When the diameter is 500 mm a typical truck can take six pipes of enclosure and conductor, which make up a total weight of 2.4 t. This value will make it on most roads.

The typical 420 kV cable is usually coiled with lengths of 300 m to 500 m per coil. With a weight of about 40 kg per metre the coil weights are in the range of 12 t to 20 t. This weight will require relatively strong streets for transportation. The weight is therefore an important project planning parameter for each project.

8.4.3   Maximum Size

The size is an important factor related to narrow roads, sharp bends or close-to-the-road housing. The overhead line towers can be separated into small-sized steel beams and the wires can be separated into small coils. The wires are relatively thin.

The cable is coiled with relatively large-sized cables to several meters in diameter for the coil. This coil needs accessibility under bridges, through tunnels or overhead lines and at railroad crossings, for example.

The GIL pipes can be separated into small sizes of, for example, 500 mm diameter for the enclosure. The length of the GIL pipes can be reduced to shorter lengths, e.g. 10 m. Short GIL section lengths require a higher total number of GIL sections and more welding.

The maximum size for accessibility of the transmission line route is an important project planning parameter.

8.4.4   Type of Soil

The soil over the route, and the size of the workshops, has a large impact on the type of equipment which can be used. Soft soils such as swamps have very different requirements from rock or clay. The greatest flexibility in terms of adapting the site works for erection of the towers and pulling the wires is given with overhead lines. The adaptation of weight and size for GIL and cables has its limitations, as explained in the paragraphs before.

The impact of soil can cause a relatively high extra cost in a transmission project. Solving the requirements of the soil is an important planning task for each project.

8.4.5   Transport Roads

The wide-area transport road or railway systems may give restrictions on size, weight or length of each element to be transported from its manufacturing place to the site. The overhead line elements will meet most of the road restrictions without big changes to the manufacturing and assembly process. There will be an impact on cost if the transportation units are small.

The cables are usually manufactured in large cable factories and then transported in coils to the site. These coils should have as many metres of cable as possible to reduce the number of joints on-site. The cable factory can hardly be relocated for a project.

The GIL is manufactured in pipe welding or extruding machines and the insulators in factories. The insulators have a small size and can easily be transported. The extruded pipes are manufactured in large factories with large extruders and are unlikely to be relocated to the site for one project. The spiral welding pipe machine can be transported and reassembled on-site within some weeks. This gives the possibility with GIL of producing the pipe on-site and reducing the transportation.

8.4.6   Space for Workshop On-Site

The on-site workshops are temporarily needed for the erection or laying of the transmission line. The overhead lines do not need much space for workshops on-site and there will be no restriction in most cases. The cable on-site works are mainly jointing of the cable segments of 300 m to 500 m. A specific working place for the joints is needed on-site for some weeks.

The GIL will mainly be assembled on-site and therefore an assembly workshop is needed. This workshop usually has double the length of one pipe segment and will be 30–40 m long with a width of about 20 m. The assembly workshop is usually a tent close to the site for laying the GIL. It will be used for two to three months per kilometre of GIL system. The assembly workshop will then be transported to the next location. If aluminium pipes are manufactured on-site a second tent of about the same size is needed.

Space requirements on-site and close to the route are part of the project planning and have an impact on the project assembly process and cost.

8.5   Soft Parameters

8.5.1   General

To get permission to build a new transmission line is not an easy task. In Europe, North America and parts of Asia, opposition by the public has a major impact on the success or otherwise of transmission line projects. Soft parameters sometimes have values but very often they are related more to emotions and feelings and can be used by politicians and public initiatives for their own goals. In most cases these goals end with the final decision.

8.5.2   Aesthetics

What is nice? What is beautiful?

Each person has his own opinion. Some feel that overhead lines are nice structures, others are strictly against overhead lines, but may accept windmills. Most people would accept overhead lines, but not in their backyard.

This is not new, but the opinion of the public has an important role today and maybe a more important role in the future. When asking for aesthetics of transmission lines, the best solution is that they are not visible.

The non-visible transmission technologies of GIL and cables need trenches underground to be non-visible. These excavations of trenches – even if only temporary – might also cause a discussion on aesthetics because after the work is done there will be a change in the landscape. This change may cause discussions on project routings.

Anyhow, from an aesthetic point of view the underground solution has more chance of getting permitted than the overhead line.

8.5.3   Non-visibility

In a permission process with public and authorities, the principle "Out of sight, out of mind!" might be a powerful argument. Compared with the gas pipeline network, which is about the same size as the electric transmission network, the discussion on overhead lines is much more intensive. Gas pipelines transport natural gas for thousands of kilometres and are underground, not visible.

For new power transmission lines the public discussion on visibility is strong and has a significant impact on the project planning and its permission.

8.5.4   Noises

There are two types of noise exposure with power transmission. One is the corona discharge of high-voltage overhead lines under certain weather conditions. The other is the noise of machines and equipment during erection.

The noise coming from corona discharge is related to overhead lines only. This noise occurs under certain weather conditions and can be heard close to overhead lines. This might cause opposition in the public when a populated area is close by.

The noises coming from the erection and laying of underground cables are mainly related to the excavation works, where large volumes of soil need to be moved. First moved away from the trench and stored. Then recycled and filtered before being brought back to close the trench. This process will last several months for each section of the transmission line.

These noises and the handling of transportation traffic will have a major impact on the project planning and cost.

8.6 Economics

The most important parameter for the success of a technology is the cost factor. GIL has been redesigned for long-length underground applications using gas mixture and pipeline laying technology, which brought the cost down by 50%. This makes the future application more likely. Also, cable and overhead line technology is constantly being improved to bring down the investment cost.

Compared with overhead lines the underground technologies of GIL and cables are more expensive at voltage levels of 420 kV and 550 kV. When higher voltage levels of 800 kV or 1000 kV are required, the cost of the GIL solution will come closer to the cost of overhead lines. This is easy to understand, because the 420 kV or 550 kV overhead line tower has a height of 30–35 m with a wire bundle up to 4 wires. The 800 kV or 1000 kV tower has a height of 60–80 m and a wire bundle of 8–10 wires to meet the requirement of maximum field strength. The GIL will only increase the 500 mm pipe of 420 kV and 550 kV systems to 630 mm pipes for 800 kV and 1000 kV systems. This is a small increase by doubling the voltage and with this the power transmission capacity.

The driver world-wide for 800 kV and 1000 kV power transmission systems is coming from the regenerative energy generation with large resources offshore and in deserts [118]. The cost relationship between different transmission systems can only be given a rough orientation and varies from project to project.

A 420 kV or 550 kV GIL has a cost factor of about 6–8 times more expensive than overhead lines. For 800 kV GIL this cost factor is down to 3–4 times and for 1000 kV a factor of 2–3 times can be expected. But this only compares the initial investment cost and not the advantages of operation in terms of losses, reliability, availability, maintenance and power ratings.

At 420 kV level for cables and GIL the cost relationship is defined mainly by the power rating. A GIL has the cost advantage if instead of one cable system, two are needed. This value is around 1800 MVA per system.

9

Power Transmission Pipeline

Offshore wind farms are under development world-wide with large projects in Europe, mainly in the North Sea. The total wind energy potential is larger than the total need for electrical energy in Europe [198, 200]. The problem is that not all regions are easily accessible, and costs for erection can be high. In the past, offshore wind farms were located close to the coast line (e.g., at 5–10 km distance). Today and in the future, offshore wind farms will be built 50–200 km away from the coast line. The connecting technology used is AC cables below 60 km and DC converter stations and DC cables when longer distances are needed [24, 150, 153, 160, 187, 188].

The use of GIL for connecting offshore wind farms offers a solution when large amounts of wind energy (e.g., 8 GW) are collected offshore and then connected by a GIL tunnel system called Power Transmission Pipeline (PTP[TM])[1] on land [171]. The EU-sponsored study has been carried out by ForWind, a competence centre at the University of Oldenburg [175] and Hannover [189], the consulting company for tunnel and pipeline aspects of ILF Beratende Ingenieure, München [190] and Siemens Energy Transmission and Distribution, Erlangen [191] – all from Germany. The sponsors of the study are the European Commission for Trans-European Networks Electricity in Brussels [192] and the Bundesministerium für Wirtschaft in Berlin, as a National Body [193].

Figure 9.1 shows one of the first large offshore wind farms erected in the North Sea, about 10–20 km away from the Danish coast line. The energy is collected at the offshore platform with 33 kV cables and then connected by a 130–150 kV AC sea cable to the grid connection point on land. Figure 9.2 shows the collecting platform with the 33 kV AC cables coming from the 2.3 MW wind turbines, the transformers from 33 kV to 132 kV, medium and high-voltage switchgear and the 132 kV AC cable for the connection of the platform to the on-land grid connection point.

The offshore wind turbine technology was developed on the basis of the on-land technology. Using the well-proven three-propeller version, the main challenge was to solve the offshore conditions with strong winds and salty air. After some experience with corrosion in the gear and bearings, the wind turbines improved quickly and solved the problems. Today, offshore wind farms are seen as reliable and can be installed offshore even far away from the coast line.

---

[1] PTP[TM] is a trade mark of Siemens and stands for Power Transmission Pipeline.

Gas-Insulated Transmission Lines (GIL), First Edition. Hermann Koch.
© 2012 John Wiley & Sons, Ltd. Published 2012 by John Wiley & Sons, Ltd.

Figure 9.1   Offshore wind farm, Horns Rev, Denmark: 2.3 MW wind turbines. Reproduced by permission of © Siemens AG

During recent years hundreds of wind turbines have been installed successfully and hundreds will follow. Turbine power is going up from 2.3 MW to 5 and 6 MW at Alpha Ventus, to 7 MW in UK offshore applications and maybe 10 MW per turbine in a few years' time.

The wind turbines combine top aeronautic knowledge of propellers with new technology in generation and transmission. Today's wind turbines installed offshore contain the top technologies, see Figure 9.3.

The bottleneck of connecting large-scale offshore wind farms to the grid on land is the landing point. Wind farm sizes have grown from 50 MW to 500 MW today and to 1000 MW in the coming years in one wind farm. The North Sea offers more than 300 GW wind capacity, which will need hundreds of landing points to get the energy on land. To concentrate the wind energy offshore and then use a strong AC GIL in a tunnel system is the idea of the PTP™). A GIL in a tunnel is shown in Figure 9.4.

9.1 Feasibility Study

The "Feasibility Study of a GIL Technology Based North Sea Network of European Offshore Wind Power Stations for Electricity Trading" in the sector of the Trans-European Networks

Figure 9.2 Offshore wind farm, Lillegrund, Denmark: offshore collecting platform. Reproduced by permission of © Siemens AG

Figure 9.3 Wind engine with 2.3 MVA power. Reproduced by permission of © Siemens AG

Figure 9.4   GIL in a tunnel. Reproduced by permission of © Alpiq Suisse SA

(TEN) [118] deals with the following topics:

- Offshore wind energy developments in Northern Europe (analyses of wind power potential, necessary transmission capacities).
- Transmission of offshore wind energy via PTP™ (adaptation of GIL for offshore usage, laying techniques).
- Integration of large amounts of wind power (connection possibilities, modelling of bottlenecks and grid overloads).
- Economic viability and environmental aspects.

The project duration of the feasibility study is from 10/2006 until 09/2009.

The development of European offshore wind energy poses a great challenge to the expansion of power grids. Several Gigawatts of electric power must be transported from wind parks on the high seas to the grid on land. As some planned wind parks lie more than 100 km off the coast, new approaches are necessary for connecting them to the grid. The positive experience gained from onshore usage of GIL, but also the need for research with respect to offshore demands, became the motivating factor behind the initiation of an EU feasibility study. PTP™ is a future-oriented new product, which is based on approaches from oil and gas pipeline technology using GIL as a medium of transport. PTP™ is understood as a total system for transporting electricity. The system includes collecting platforms for the feed-in of electricity from different offshore wind parks as well as the laying technology for the transmission system via tunnel construction. GIL is the component of PTP™ used for the transport of electricity. PTP™ is the offshore enhancement of GIL.

Table 9.1   Nominal capacities of offshore wind farms in the North Sea (as at 10/2008)

Country	Total	In operation	Approved	Applied for	Under construction	Rejected	Unspecified
Belgium	1 246	0	330	0	300	200	416
Denmark	360	160	0	0	200	0	0
Germany	62 762	9	5 407	20 637	60	0	36 649
France	195	0	105	0	0	0	90
The Netherlands	25 285	228	0	24 476	0	581	0
Great Britain	6 115	340	1 500	340	472	108	3 355
Total	**95 963**	**737**	**7 342**	**45 453**	**1 032**	**889**	**40 510**

(Power in MW)

9.2   Offshore Wind Energy in Europe

The European Union wants to reduce CO_2 emissions by 20% by the year 2020. In doing so, the EU is placing particular emphasis on the expansion of offshore wind energy. Currently, the development of offshore projects has advanced to the furthest stages in Denmark, Great Britain, the Netherlands and Sweden. Above all, offshore wind energy is to be expanded in the North Sea. In the long term, very high power densities are to be expected off the coasts of the Netherlands, Great Britain and Germany (Table 9.1). Offshore wind park projects with a total power of over 7000 MW have already been approved across Europe. Further projects comprising roughly 45 450 MW of power are in the process of approval. Taken together, this corresponds to more than half of the generated gross production capacity of 98 000 MW from fossil fuels in the year 2006 [200].

The dimensions of the planned expansions show that new grid connection solutions must be acquired alongside the development and approval of offshore projects. This calls for a transmission system that is able to transport high amounts of power efficiently over long distances. In some regions of the North Sea, for example along the German coast and in the "Wadden Sea", the possibilities of laying transmission lines are spatially limited. Nature reserves, shipping and other important areas of usage such as fishing and tourism must be taken into account. In addition to the low loss and efficient transmission of power, there is also a demand to use the smallest possible area through spatial concentration. This includes environmentally sound laying methods minimizing conflicts of usage.

The principle collection of offshore wind energy is shown in Figure 9.5 when the PTP is used. From wind clusters the energy is concentrated and brought on land and then distributed again to the load centres.

9.3   Under Sea Tunnel System

How can high capacities of wind power generated at high sea be transported on land using the under sea tunnel system PTP^{TM}?

This can be explained using a model transmission route: the exemplary transmission route has 8000 MVA of transmission power connecting several offshore wind farms with the North Sea coast. Figure 9.5 shows a schematic PTP^{TM} transmission route with substation platforms C1, C2 and C3 and a connection to a fictional landing point. The electricity generated from

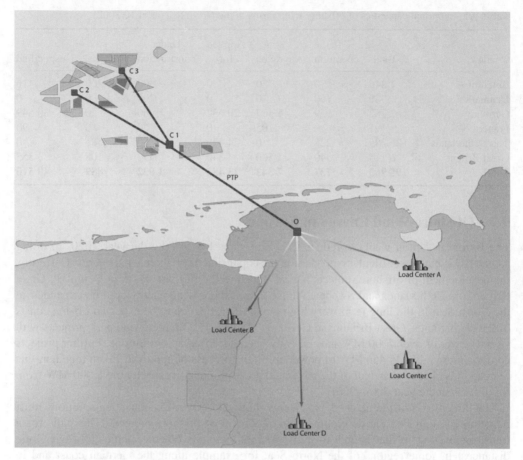

Figure 9.5 Example PTP^{TM} with collection points. Reproduced by permission of © Carl von Ossietzky Universität Oldenburg

offshore wind farms is fed into the PTP^{TM} collecting platforms using sea cables. PTP^{TM} conduit systems deliver the electricity ashore. An on-land substation (O) connects PTP^{TM} with the transmission grid. From there, the energy can be transported by means of alternate or direct current to load centres.

In detail, the model entails the following. Single wind power plants are connected within the wind farms. Wind farms with a total of 400 to 500 MW are connected by sea cables to a PTP^{TM} collecting platform. Via PTP^{TM}, the electric power is transmitted from the collecting platform to the connecting substations on the coast (O). From there it is passed on to load centres.

The core feature of the concept is a tunnel with GIL inside linking the offshore collecting platform C1 with the grid on land. In ecologically and use-sensitive areas, e.g. the protected areas of the Wadden Sea, a double tunnel is bored underneath the sea floor. Outside sensitive areas the tunnel can be laid using the immersed tunnel method. Figure 9.6 gives an overview of the concept showing the route between the substation on land and the offshore collecting platform C1.

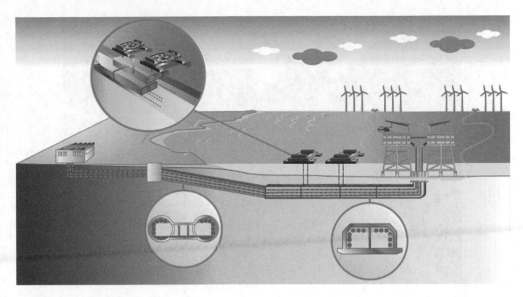

Figure 9.6 Overview of the offshore wind farm connection scheme with PTP™. Reproduced by permission of © Natterer

PTP™ offshore consists of:

- a GIL as the primary technology for energy transmission,
- secondary GIL technology such as monitoring facilities,
- constructions like the tunnel and platforms including equipment.

Why use GIL in an under sea tunnel system?

Traditionally, GIL technology is implemented in substations or in power plants, mainly in the highest voltages ranging from 400 to 800 kV. This allows the transmission of 2000–4000 MVA of electric power per three-phase system. A pressurized insulating gas mixture serves as a reliable insulation. GIL are single-phase, co-axially constructed pipes made of aluminium implemented for high-voltage alternating currents. In all, GIL consists of two pipes: an inner conductor pipe is centred within a grounded enclosure pipe. The conductor pipe is held by a few insulators made of epoxy resin. The space in-between is filled with an insulating gas mixture consisting of nitrogen (N_2) and sulphur hexafluoride (SF_6). The enclosure pipes employed are spirally welded pipes made of aluminium alloy (see Figure 9.7). Extruded pipes are used as conductor pipes. They consist of 99.9% "electrically" pure aluminium. The conductor and enclosed pipes are welded together on-site, thus forming a gas-tight casing for which a refilling of gas is unnecessary throughout the entire lifespan. The transmission route is divided into segments of approximately 1 km in length in order to limit the release of gas in the event of damage. Necessary changes in direction are accomplished with angle elements up to 90°.

Figure 9.7   GIL in a tunnel, two three-phase systems. Reproduced by permission of © Alpiq Suisse SA

Gas-insulated pipes are typically laid in a tunnel, above ground or directly buried in the ground when coated with a corrosion protection layer. Inside substations, GIL are also installed on steel structures above ground. For offshore applications, the PTP™ has been developed with two possible tunnel-laying techniques.

Electro-technical Properties of GIL   A GIL has several special electrical properties at its disposal. Transmission losses are very low due to the large cross-section of the conductor, which is important for long-distance transmission (see Figure 9.5). The solidly grounded enclosure pipe results in low magnetic fields outside the GIL. External electric fields do not exist. Operational and personnel safety is high due to the solid metal enclosure. As a passive system of transmission, GIL needs little maintenance over its very long lifespan and at a high degree of operational safety.

GIL Technical Data
- Voltage: 400 to 800 kV
- Transmission power: 2000 to 4000 MVA
- Material: aluminium pipes
- Enclosure diameter: 500 to 630 mm
- Insulating medium: gas mixture at 0.7 to 0.8 MPa
- Weight: 30 to 40 kg/m

Electro-technical properties:

- High transmission power over long distances.
- Low transmission losses due to large diameter.

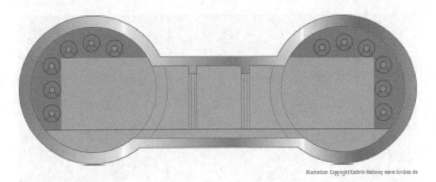

Figure 9.8 Profile of two drilled tunnel tubes with PTP™ for 8000 MVA. Reproduced by permission of © Natterer

- Low magnetic fields.
- Solidly grounded enclosure pipes.
- High degree of operational and personnel safety by metallic encapsulation.

In Figure 9.8 the cross-section of a double-tunnel system is shown, which can take four systems of GIL with a total power transmission capability of 8000 MVA.

PTP™ Laying Technique   The construction of the tunnel must be finished first. It is subsequently equipped with GIL. The tunnel features include energy supply, ventilation, monitoring and communications systems as well as a transport system. The transport system is necessary to overcome the long tunnel distances during the construction period and later for servicing purposes. Tunnels are constructed with at least two pipes for reasons of operational and personnel safety.

Tunnel Constructed with a Drift Mining Tunnel Machine   The exact approach to tunnel construction with a tunnel boring machine (TBM) depends on the geological and hydro-geological circumstances. A fundamental aspect all procedures share is that ground is dismantled gradually under the protection of a shield. The tunnel tubes are lined with waterproof concrete shell elements (tubbings, Figure 9.8). Material, devices and machines are brought into the tunnel through the tunnel portal using the transport system. Excavated material is taken out using the same route. Intermediate constructions such as shafts are not necessary using this construction technique.

Two tunnel construction methods are available:

- A closed construction in miner's fashion using a TBM, such as those used world-wide for the construction of traffic tunnels.
- The immersed tunnel method, which is a part of current offshore building techniques.

The "immersed tunnel method" consists of pre-casting individual tunnel segments made of reinforced concrete in dry or floating docks. This construction method allows the manufacture of several tunnel pipes in parallel. Individual tubes are separated by dividing individual segments lengthwise with separating walls. The tunnel elements are tugged by draggers to the

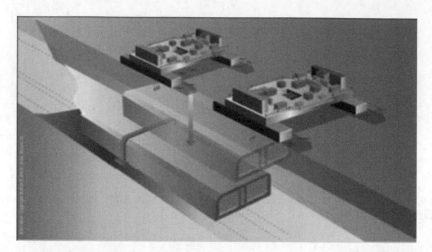

Figure 9.9   Lowering of tunnel segments from tug pontoons. Reproduced by permission of © Natterer

position where they are lowered to the seabed. The front sides are sealed for transport with water-tight compartments. At the point of destination, the tunnel segment is control-lowered into a trench on the seabed. Waterproof connections are established with elements that have already been placed. Lowering takes place by adding ballast to offset buoyancy. To this end, the insides of the individual elements have ballast tanks. By slowly filling the tanks with water, the weight of the element increases and it sinks. Tug pontoons with winches are primarily on hand to secure the segment's position (Figure 9.9). Thereafter, tunnel segments are covered with sand or gravel. By replacing the ballast water with ballast concrete and an additional cover of rocks, the tunnel is protected from the effects of buoyancy and secured against other external influences.

GIL is laid after the completion of the tunnel. Pre-assembled GIL elements are delivered to the landing shaft and lowered into the shaft using a crane. In the welding tent individual elements are completely assembled, welded together and then pulled into the tunnel. The GIL is kept on rollers or mounts equipped with sliding contact bearings, allowing thermal expansion of the coat tube. The compensator units, built in at regular intervals, permit the respective thermal expansion in the conductor tubes.

9.4   Offshore and Onshore PTP^{TM} Constructions

Offshore Collecting Platform   On the offshore PTP^{TM} collecting platform, wind energy generated offshore by 1600 wind turbines is bundled. The voltage is stepped up to 400, 500 or even 800 kV, thereby permitting the concentrated transmission of energy to the mainland via four GIL systems. The offshore collecting platform (Figure 9.10) consists of a total of three platforms: a tunnel connection platform and two substation platforms. Typical construction elements from the oil and gas industry are used for the construction of the substation platforms.

Platforms placed on steel framework constructions with pile foundations, so-called jackets, host both substations with 4000 MVA each. Each platform consists of three decks, whereby

Figure 9.10   PTP™ platform with transformer stations and facilities. Reproduced by permission of ©
Natterer

the transformers and switchgear are found on the two bottom floors. Each of these bundles
2000 MVA for the transmission via one GIL system. The top deck hosts utility services such
as accommodation for personnel, controlling and communication facilities, safety systems,
a helicopter deck and additional equipment for the operation of the platform. The tunnel
connecting platform – the connecting element between the substation platforms – is installed
on a concrete platform held in place by gravity. A vertical concrete shaft leads the GIL from
the tunnel over two steel framework bridges into the transformer platforms.

Intermediate Shaft   The transition from the drilled tunnel to the immersed tunnel requires a
handover construction. This is the intermediate shaft. The intermediate shaft can be sealed off
under water or give access above the water surface.

Landing Shaft   The PTP™ landing shaft serves as a transition for the offshore tunnel to the
PTP™ on land. For onshore laying, a buried or an open construction above ground comes into
consideration.

Main Transformer Substation   The main transformer substation distributes the bundled energy to grid feed-in points. Due to the high energy throughput, a buried GIL could be a suitable
transmission solution.

Timeframe to Completion   The planning and approval phase, the construction of the necessary
manufacturing capacities and logistics, the construction of a 60 km-long tunnel segment with
transformer platforms as well as the installation of the four PTP™ systems (8000 MVA)
require an estimated time of 10–12 years.

Platform (8000 MVA)

- Depth: 35 m
- Weight: 2×7000 t
- Jacket height: 50 m

Dimensions of platform superstructures

- Height: 25 m
- Width: 40 m
- Length: 60 m

9.5 Next-Generation Technology

Directly laying aluminium GIL in a seawater environment requires steel casing pipes due to the properties of the material. In the case of seawater laying, a protective casing pipe is used to prevent the GIL from damage. The GIL aluminium pipe is therefore inserted into a steel pipe for protection. As with offshore pipelines from the oil and gas industry, the laying takes place on offshore pipe-laying vessels. Here, the entire assembly of GIL and casing pipes is carried out aboard the ship. During laying, the tough requirements regarding external conditions during construction stand in the forefront. In addition to procedures for laying, welding and connecting the various pipes, material properties must be taken into account during construction. As the laying, operational and repair technology of this extremely complex pipe-in-pipe-in-pipe system will be required for the foreseeable future, this concept is currently considered as a future option (see Figure 9.11).

9.6 Offshore Environment

The North Sea is a space that is intensively used and occupied. It ranks among the most heavily travelled transport routes for goods and it is a region of intensive natural resource mining (oil and gas production, sediment extraction) as well as other economic activities such as fishing,

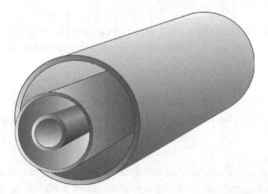

Figure 9.11 Pipe-in-pipe-in-pipe system for offshore laying. Reproduced by permission of © Carl von Ossietzky Universität Oldenburg

mariculture and tourism. This automatically leads to competition with offshore wind energy developments as there is only a limited amount of sea area available.

Encroachments upon the underwater world lead to further conflicts, particularly regarding ecologically sensitive and especially protected areas like the Wadden Sea, national parks and Nature 2000 regions. Here, particular national and international guidelines for the protection of the environment are in effect, which also play a significant role for offshore wind park planning:

- The Flora-Fauna-Habitat Directive (FFH-RL: 92/43/EWG) [11].
- The Bird Protection Directive (V-RL: 79/409/EWG) [194].
- The future Marine Strategy Framework Directive of the EU.
- National provisions.

For this reason, grid integration concepts avoiding or reducing the level of encroachment on the Wadden Sea are especially important for an environmentally and legally sound planning. In regions where large amounts of power are generated and transmitted but where the spatial dimensions of transmission routes are limited due to environmental and other constraints, PTP^{TM} transmission routes can offer a useful solution. With a high capacity of 2000 MW per GIL system and up to 8000 MW per PTP^{TM} route, energy can be bundled and transmitted requiring few high-power routes.

In areas near the coast and in those which are especially sensitive (e.g., the Wadden Sea national park), the laying of PTP^{TM} using the tunnel-boring method presents itself as a solution. On land, there is merely a need for access to the insertion of a tunnel-drilling machine. This will hardly be visible after finishing the construction. In the model transmission route introduced in Chapter 1, the first offshore collecting platform situated 60 km off the coast and the East Frisian islands is not visible from shore. The maritime picture of the landscape will not be impaired.

Outside sensitive areas, the tunnel can be continued by lowering pre-built tunnel elements (immersed method). Large laying ships are necessary to transport the required materials from the harbour to the point of destination. Depending on the subsoil, sediment can be swirled up while flushing the tunnel channel, lowering and pile jetting segments, which in turn leads to fish and marine mammals being temporarily chased away from the areas of construction. To the same degree, vibrations and noise emissions can lead to the impairment of fish and mammals (through shipping traffic and near substation foundations). After the completion of the laying construction, the affected habitats will be re-populated by fish and mammals.

Sediment movements by transport take place in various ways. With the tunnel-boring method the excavated material is transported via vehicles placed on tracks, while the immersed tunnel method relies on ships to load the sediments.

10

Conclusion

Gas-Insulated Transmission Lines (GIL) are a high-power transmission technology which can be installed above ground, in a trench or tunnel underground, or directly buried into the ground. Invented in the 1960s, the first industrial application was built as a 400 kV power transmission system in a hydropower storage plant at Schluchsee in Germany in 1974. Since then more than 300 km phase length has been installed and is in operation world-wide. After this first step in the invention of GIL, a second step followed in 2001 with the first installation using N_2/SF_6 gas mixture technology and welding for jointing on site at the PALEXPO project in Geneva, Switzerland. Laid in a tunnel, this installation identifies the start of the second-generation GIL for long-distance applications. In 2010 the first directly buried second-generation GIL followed, with the Kelsterbach project at the Frankfurt International Airport.

Today, GIL is applied in many projects around the world when high-power transmission with high reliability and maximum availability is required. The sizes of the projects are constantly increasing: from typically some hundred metres system length in the past to typically several kilometres today. This is a factor of 10 times longer. And in the future, even longer-length transmission lines of some 10 km will be realized under the conditions of regenerative power.

The GIL technology is based on gas insulation with cast resin insulators and aluminium housing. This simple design has proven very reliable over the last 35 years of operation. Design and routine test and a proven quality insurance system for on-site assemblies are the basis of this high-level reliability.

From all installations in the world (more than 300 km phase length), no critical failure has been reported over the last 35 years. An excellent revenue in reliability.

The different laying options are above ground on steel structures, in a tunnel or trench, or directly buried – offering a large field of possible applications inside substations and power plants or even outside as cross-country installations. The high safety level of high-voltage GIL allows us to add GIL into traffic tunnels together with railways or road traffic. Also, adding GIL to traffic bridges is possible and – because of the solid outer enclosure – the GIL is easy to install or fix in a tunnel or on a bridge.

Long-duration testing of GIL for tunnel-laid and directly buried versions has given important proof of the practicability on-site in large-scale projects, and information about the excellent long-time stability of the gas-insulated transmission line for an expected lifetime of 50 years.

Gas-Insulated Transmission Lines (GIL), First Edition. Hermann Koch.
© 2012 John Wiley & Sons, Ltd. Published 2012 by John Wiley & Sons, Ltd.

From a system and network point of view, the GIL can be fitted into the high-power transmission network of voltage levels typically 400 kV and higher. With the availability of 500 kV, 800 kV and even 1200 kV GIL solutions, the upper sector of high-power transmission can be covered. To transmit high power ratings over long distances, extra high voltage levels are the best solution.

From an environmental point of view, GIL offers several advantages of being invisible under ground, having lower transmission losses, with very low electromagnetic fields and a high safety level concerning surrounding equipment or nearby personnel. The correct handling of SF_6 used as an insulating gas, which should not be released to the atmosphere, is an important operational issue. From the installation point of view the completely welded design of the second-generation GIL is gas-tight. The proper gas handling is regulated by international standards and must be followed. This reduces the amount of gas released to the atmosphere to a negligible minimum.

In comparison with other transmission systems the main advantage of the GIL is power transmission capability over long distances. Simply said: "The higher the power rating and voltage the better for GIL!"

Power ratings at 400 kV level of 2000 MVA per electrical three-phase system, or 3000 MVA for 500 kV, or 4000 MVA for 800 kV or 6000 MVA for 1000 kV are typical values. The low capacitive values of gas-insulated systems also allow long lengths without the need for compensation. At 400 kV a typical length is in the range of 60–80 km depending on the local network conditions.

New fields of future application of GIL are in combining road and railway tunnels or bridges with GIL. Also, it is possible to combine high-power transmission with highways or railroad. Or, as concentrated high-power underground electric in a small corridor. The connection of large offshore wind farms and the landing of electrical energy at a small slot on the coast is a possible GIL application in the future.

After 35 years of very positive experiences with GIL, this technology is now ready to solve the high-power transmission requirements of the future.

References

[1]   George G., Interconnecting the globe, *Transmission and Distribution World*, 01/2000.
[2]   Li F., Haubrich H.-J., Bewertung von Netzausbaumaßnahmen zur Engpassbeseitigung im UTCE-Verbundnetz, *Energiewirtschaftliche Tagesfragen*, 56. Jg., 2006.
[3]   Giebel G., Nielson H., Hurley B., Bigger transmission distance v lower load factors: the European dilemma, *Modern Power System*, 10/2005.
[4]   Rajgor G., Big Hike upwards for five year forecast, *Windpower Monthly*, 05/2008.
[5]   Koch H., Underground cables and gas insulated transmission lines, *Wiley Encyclopedia of Electrical and Electronics Engineering* (Webster), 2006.
[6]   Koch H., *Gasisolierte Übertragungsleitungen (GIL) für hohe Übertragungsleistungen*, ETG-Tagung, Zürich, Schweiz, 05/2000.
[7]   Koch H., Kumar A., Christl N., Lei X., Povh D., Retzmann D., Advanced Technologies for Power Transmission and Distribution – Benefits and Impact of Innovations, Siemens Brochure, 2010.
[8]   Klein J., Hochspannung im Rohr – Neue Übertragungsleitungen revolutionieren die Stromversorgung, DMT-Innovativ spezial, 09/2001.
[9]   Koch H., Gas-insulated transmission line (GIL), in J. McDonald (ed.), *Power Substations Engineering*, CRC Press, Boca Raton, FL, 2003.
[10] Koch H., Pöhler S., Schmidt S., Anwendungsvorteile von gasisolierten Übertragungsleitungen (GIL) für unterirdischen Energietransport in Ballungszentren, etz 1-2/2002.
[11] Flora-Fauna-Habitat Directive (FFH-RL: 92/43/EWG). http://ec.europa.eu/environment/nature/legislation/habitatsdirective/index_en.htm
[12] Zimmer C., Eckenroth L., Feldmann J., Jones W., Elms N., Riechmann C., Schlecht D., Cremer D., Haubrich H.-J., Weiterentwicklung des grenzüberschreitenden Engpassmanagements im europäischen Stromnetz, *Energiewirtschaftl. Tagesfragen*, 54. Jg., Heft 12, 2004.
[13] Pöhler S., Koch H., Anwendungsmöglichkeiten der GIL im europäischen Verbundnetz, ETG von Electrosuisse, Fribourg, Schweiz, 01/2004.
[14] Koch H., Auswirkungen der gasisolierten Leitungen (GIL) auf das Energienetz, ETG-Kongress 2003, Hamburg.
[15] http://de.wikipedia.org/wiki/Cahora_Bassa August 2011.
[16] http://de.wikipedia.org/wiki/Itaip%C3%BA August 2011.
[17] http://en.wikipedia.org/wiki/Pacific_DC_Intertie August 2011.
[18] Liu Q., Lu Q., Rehtanz C., Vereinigte Netze, ABB Technik, 02/2005.
[19] http://cn.siemens.com/cms/cn/English/PTD/products/references/utility/Pages/Gui-Guang.aspx
[20] http://www.ptd.siemens.de/artikel0605.pdf
[21] Povh D., Pyc I., Retzmann D., Weinhold, Future Developments in Power Industry, The 4th IERE General Meeting and IERE Central and Eastern Europe Forum, Krakow, Poland, 10/2004.
[22] Povh D., Retzmann D., Rittiger J., Benefits of Simulation for Operation of Large Power Systems and System Interconnections, IERE General Meeting, San José, Costa Rica, 11/2003.

[23] Koch H., Innovative Trends in the High Voltage Field, IEEE PES Substations Committee Meeting, Bellevue, WA, 04/2007.

[24] Lange B., Hahn B., Offshore-Testfeld Borkum, Nachhaltig Wirtschaften, 01/2007.

[25] Koch H., To solve bottle-necks in the European Transmission Net, IASTED, Benalmádena, Spain, 06/2005.

[26] Koch H., AC Bulk Power Systems in Metropolitan Areas Application, IEEE/PES T&D Asia Pacific, Dalian, China, 08/2005.

[27] Koch H., Schoeffner, G., Gas-Insulated Transmission Line – To Solve Transmission Tasks of the Future, IPEC Conference, Singapore, 05/2003.

[28] Koch H., Gas-Insulated Transmission Line (GIL), Chapter 18 of *Electric Power Substations Engineering*, 05/2003.

[29] Jahnke B., Hanse S., Energy transmission in Ballungsräumen mit Hoechstspannungskabeln bis 400 kV, *Elektrizitätswirtschaft*, 12/1992.

[30] Pedersen W., Doepke K., Brolin R., Development of a compressed gas-insulated transmission line, *IEEE Power*, 1970.

[31] Abilgaard E., Zur Verlegung SF6-isolierter Hochspannungsrohrkabel, Sonderdruck aus ENERGIE und TECH-NIK, 1972.

[32] Koch H., Magnetic Fields of Gas-Insulated Transmission Lines – Calculations and Measurements, ISH Conference 2003, Delft.

[33] Koch H., Hillers T., Gasisolierte Hochspannungsleitungen erschließen neue Wege der Energieübertragung, ETZ, Heft 11, 1997.

[34] Chakir A., Koch H., Gas Insulated Transmission Lines for High Power Transmission over Long Distances, UPDEA, Senegal, 04/99.

[35] Bär G., Dürschner R., Koch H., 25 Jahre Betriebserfahrung mit GIL – heutige Anwendungsmöglichkeiten, etz, Heft 1-2/2002.

[36] Hillers T., Koch H., Gas Insulated Transmission Lines (GIL): A solution for the power supply of metropolitan areas, CEPSI, Thailand, 07/1998.

[37] Koch H., Erste qualifizierte gasisolierte Übertragungsleitung für Mischgas, ev-report, 03/1999.

[38] Henningsen C.G., Kaul G., Koch H., Schütte A., Plath R., Electrical and Mechanical Long-Time Behaviour of Gas-Insulated Transmission Lines, CIGRE Session 2000, Paris.

[39] Pedersen W., Doepken K., Brolin R, Development of Compressed Gas-Insulated Transmission Line, 03/1995.

[40] Thuries E., Girodet A., Roussel B., Guillen, Underground Gas-Insulated Transmission Line, Villeurbanne, *AEEE Journal* Vol. 7 08/1996.

[41] Thuries E., Voisin G., 420–550 kV Gas Insulated Transmission Lines, Gec Alsthom Technical Review, 1994.

[42] Schoeffner G., Neumann T., Application of Gas Insulated Transmission Lines (GIL) and Gas Insulated Switchgear (GIS) for Power Plants.

[43] Koch H., Schoeffner G., Gas-Insulated Transmission Line – To Solve Transmission Tasks of the Future, IPEC Conference 2003, Singapore.

[44] Koch H., Siemens Activities on the Development and Application of GIL, CIGRE SC23, Colloquium Venezuela 10/2001.

[45] Koch H., Gas-Insulated Line (GIL) of the 2nd Generation, IEEE Conference AC/DC Power Transmission, 2001.

[46] Koch H., Gasisolierte Übertragungsleitungen der 2ten Generation für das Hochspannungsnetz, e&i, Österreich, Heft 1, Jg. 119, 01/2002.

[47] Koch H., Development, long duration testing and first application of Gas-Insulated Transmission Line (GIL), IEEE ICC Fall Meeting, 10/2000.

[48] Siemens verwirklicht in Genf weltweit erste Hochspannungsverbindung mit gasisolierter Übertragungsleitung der zweiten Generation, Pressereferat EV, Erlangen, Germany, 2001.

[49] Alter J., Koch H., Gasisolierte Übertragungsleitung (GIL) der zweiten Generation, elektrizitätswirtschaft, Jg. 100, Heft 8, 04/2001.

[50] Koch H., Pöhler S., Schmidt S., Vorteile der Energieübertragung mit gasisolierten Leitungen (GIL), etz, Heft 6/2002.

[51] IEC 62271-4: High-voltage switchgear and controlgear – Part 4: Use and handling of sulphur hexafluoride (SF6), to be published 2012, IEC, Geneva, Switzerland.

[52] Chakir A., Koch H., Numerische Untersuchung der turbulenten, natürlichen Konvektion in horizontalen, koaxialen Zylindern mit CFD Flotran-Ansys, CAD-FEM Users' Meeting, Bad Neuenahr, 10/1998.

[53] IEC 60287 Electric cables – calculation of the current rating, IEC, Geneva, Switzerland, 2006.

[54] IEC 62271-204. High-voltage switchgear and controlgear – Part 204: Rigid gas-insulated transmission lines for rated voltage above 52 kV, IEC, Geneva, Switzerland, 2011.

[55] Koch H., Schoeffner G., Gas-Insulated Transmission Line (GIL) – An Overview, ELECTRA, 12/2003.

[56] Eidinger, A., Aufbau, Einsatz und Erprobung von SF6 Rohrgaskabel, Patent DE 19604485A1 Brown Boveri AG, Baden, Argau, Switzerland, 08/1997.

[57] Schichler U., Diessner A., Gorablenkow J., Dielectric on-site testing of GIL, Proceedings of the 7th International Conference on Properties and Applications of Dielectric Materials, Nagoya, Japan, June 1–5, 2003.

[58] IEC/TR 62271-303 ed1.0 (2008-07). High-voltage switchgear and controlgear – Part 303: Use and handling of sulphur hexafluoride (SF_6), IEC, Geneva, Switzerland, 2008.

[59] Diessner A., Finkel M., Grund A., Kynast E., Dielectric properties of N_2/SF_6 mixtures for use in GIS or GIL, High Voltage Engineering Symposium, August 22–27, Conference Publication No. 467, © IEE, 1999.

[60] Koch H., Preferential Subject D1 PS2 Q6: Is there any experience from the application in the field of N_2/SF_6 mixtures and especially in high voltage Gas-Insulated Lines?, CIGRE Session 2004, Paris, 08/2004.

[61] Alter J., Ammann M., Boeck W., Degen W., Diessner A., Koch H., Renaud F., Pöhler S., N_2/SF_6 gas-insulated line of a new GIL generation in service, CIGRE Session, Paris, 08/2002.

[62] van Brunt C., SF_6/N_2 mixtures basic and HV insulation properties, *IEEE Transactions on Dielectrics and Electrical Insulation*, 10/1995.

[63] van Brunt C., SF_6 Insulation; possible greenhouse problems and solutions, Electricity Division, DEIS, 1996.

[64] van Brunt C., Olthoff J.K., Green, Gases for Electrical Insulation and Arc Interruption: Possible Present and Future Alternatives to Pure SF_6, National Instutute of Standards and Technology (NIST), U.S. Department of Commerce, Washington, USA, 11/1997.

[65] Marsden S., Hopkins M., Eck C.R., Lightning impulse withstand performance of a practical GIC with 5 and 10 percent SF_6/N_2 mixtures, IEEE, Electrical Insulation, Conference Record, 06/1998.

[66] Zende C., Pfeiffer W., Schoen D., Corona stabilisation and prebreakdown development in SF_6 inhomogeneous fields stressed with very fast transient voltages, 10th International Symposium on High Voltage Engineering, Montreal, Canada, 1997.

[67] Hairour M., Bessede J.-L., Girodet A., Toureille A., Agnel S., Meijer S., Dielectric withstand of N_2, CO_2 in GIS, CEPSI, India, 2006.

[68] Hoshina Y., Sato M., Murase H., Toyoda M., Kobayashi A., Dielectric properties of SF_6/N_2 gas mixtures on a full scale model of the gas-insulated busbar, IEEE, PES, Winter Power Meeting, New York, 2000.

[69] Hoshina Y., Sato M., Murase H., Aoyagi K., Hanai M., Kaneko E., The surface flashover characteristics of spacer for GIS in SF_6 gas and mixtures, *Gaseous Dielectrics* Vol. VIII, Plenum Press, New York, 1998.

[70] Boeck W., Taschner W., Luxa G., Menten L., Insulating behaviour of SF_6 without solid insulation in case of fast transients, CIGRE Session Paris, France, 1986.

[71] Braun J.-M., Addis G., Diederich A., Girodet A., Long-term reliability of cast epoxy insulators in gas insulated equipment, CIGRE Session Paris, France, 1990.

[72] Dießner A., Luxa G., Neyer W., Elektrische Langzeitversuche an Epoxidharz- Isolatoren in SF_6-isolierten metallgekapselten Schaltanlagen, Siemens-Zeitschrift, 1986.

[73] Knobloch H., The comparison of arc-extinguishing capability of sulfur hexafluoride (SF_6) with alternative gases in high-voltage circuit-breakers, *Gaseous Dielectrics* VIII, International Symposium on Gaseous Dielectrics, Virginia Beach, VA, June 2-5, 1998.

[74] Koch H., Experience with 2nd generation gas-insulated transmission line GIL, JICABLE Conference, Versailles, France, 06/2003.

[75] Dießner A., Prinzipien und Prüfkonzeption für gasisolierte Übertragungssysteme (GIL), Highvolt Kolloquium, Cottbus, Germany, 05/1999.

[76] Koch H., Schuette A., Gas-Insulated Transmission Lines (GIL), type tests and prequalification, JICABLE, Versailles, France, 06/1999.

[77] Kynast E., Koch H., Hochspannungstechnische Gesichtspunkte zur Entwicklung, Prüfung und zum Betrieb von gasisolierten Übertragungssystemen, Micafil Symposium, Stuttgart, Germany, 09/1999.

[78] Chakir A., Koch H., Transient thermal behavior of buried Gas-Insulated Transmission Lines (GIL) at variable loads, *ASME Journals*, Proceedings of the ASME International Mechanical Engineering Congress and Exhibition, New York, USA, 11/2001.

[79] Hillers T., Koch H., Gas-Insulated Transmission Lines (GIL): The solution for high power transmission underground, IEEE, PES Power Delivery Conference, Madrid, Spain, 08/1999.

[80] Chakravorti S., Steinbigler H., Capacitive-resistive field calculation on HV bushings using the boundary-element method, IEEE Transactions on Dielectrics and Electrical Insulation, 04/1998.

[81] Koch H., Long lengths high power gas-insulated high-voltage transmission line, *CIGRE Symposium*, Shanghai, China, 2003.

[82] IEC 62271-1 ed1.0 High-voltage switchgear and controlgear – Part 1: Common specifications, IEC, Geneva, Switzerland, 2007-10.

[83] EN 50052 Cast aluminium alloy enclosures for gas-filled high-voltage switchgear and controlgear, IEC, Geneva, Switzerland, 1986.

[84] EN 50064 Wrought aluminium and aluminium alloy enclosures for gas-filled high-voltage switchgear and controlgear, IEC, Geneva, Switzerland, 1989.

[85] EN 50068 Wrought steel enclosures for gas-filled high-voltage switchgear and controlgear, IEC, Geneva, Switzerland 1991.

[86] EN 50069 Welded composite enclosures of cast and wrought aluminium alloys for gas-filled high-voltage switchgear and controlgear, IEC, Geneva, Switzerland 1991.

[87] EN 50089 Cast resin partitions for metal enclosed gas-filled high-voltage switchgear and controlgear, IEC, Geneva, Switzerland 1992.

[88] EN 50187 Gas-filled compartments for a.c. switchgear and controlgear for rated voltages above 1 kV and up to and including 52 kV, IEC, Geneva, Switzerland 1996.

[89] Babusci B., Colombo E., Speziali, Assessment of the behaviour of gas-insulated electrical components in the presence of an internal arc, ENEL Research, CIGRE Session, Paris, France, 1998.

[90] Chakir A., Koch H., Thermal calculation for buried Gas-Insulated Transmission Lines (GIL) and XLPE-cable, IEEE Winter Power Meeting, Columbus, OH, 01/2001.

[91] Völcker O., Koch H., Insulation co-ordination for gas-insulated transmission lines (GIL), *IEEE Transactions on Power Delivery*, PE-102 PRD, 07/2000.

[92] Holmberg H.-E., Gubanski S.-M., Motion of metallic particles in gas insulated systems, *IEEE Electrical Insulation Magazine*, 08/1998.

[93] Chakir A., Koch H., FE-Berechnung der mechanischen Beanspruchungen von erdverlegten, gasisolierten Übertragungsleitungen (GIL), CAD-FEM Users' Meeting, Sonthofen, Germany, 10/1999.

[94] Morcos M.-M., Manhatten K.-S., Srivastava K.-D., On electrostatic trapping of particle contamination in GIL systems, *IEEE Transactions on Power Delivery*, 10/1992.

[95] Dale S.-J., Particle trap with dielectric barrier for use in gas insulated transmission lines, Westinghouse Electric Corp., US Patent 4335268, 06/1982.

[96] Chakir A., Koch H., Turbulent natural convection and thermal behaviour of cylindrical Gas-Insulated Transmission Lines (GIL), IEEE PES Summer Meeting, Vancouver, Canada, 07/2001.

[97] Metwally I.A., Thermal and magnetic analyses of gas-insulated lines, *International Journal of Thermal Sciences*, THESCI-D-0800364, 2008.

[98] Chakir A., Koch H., Numerical solution for a turbulent natural convection in cylindrical horizontal annuli, *ASME Journal of Heat Transfer, American Society of Mechanical Engineers*, 2000.

[99] Chakir A., Koch H., Seismic calculations of directly buried gas-insulated transmission lines (GIL), IEEE/PES T&D Conference, Asia Pacific, Yokohama, Japan, 10/2002.

[100] Koch H., GIL as a modern technology to transmit electrical power, International Power Engineering Conference, Singapure, 05/2001.

[101] Poehler S., Gasisolierte Leitungen – bewährte Technik für umweltfreundliche Unterflurverlegung von Freileitungend in sensibler Landschaft, Pressekonferenz Innsbruck, Austria, 06/2001.

[102] Kaul G., Waller R., Die gasisolierte Übertragungsleitung in der Entwicklung, *Elektrizitätswirtschaft*, Jg. 99, Heft 11, 06/2000.

[103] Hennigsen, C.-G., Polser, K., Obst, D.; Berlin Creates 380 kV Connection with Europe, Transmission and Distribution World Journal, July 1998.

[104] Engelhard G., Hillers T., Anwendung des Orbital-Reibrührschweißen von Aluminiumrohren, Schweißtechnische Tagung, Kassel, Germany, 09/2002.

[105] Engelhard G., Hillers T., Orbital-Reibrührschweißen von Aluminiumrohren, Schweißtechnische Tagung, Kassel, Germany, 09/2002.

[106] Chakir A., Koch H., Corrosion protection for gas-insulated transmission lines, IEEE PEC Summer Meeting, Chicago, USA, 07/2002.

[107] Koch H., Gas-insulated Transmission Lines (GIL) for high power transmission, FIILE 2001 Conference, Paris, 03/2001.

[108] Schöffner G., Becks M., Ollständiger Korrosionsschutz für gasisolierte Übertragungsleitungen, etz Heft 6/2003.

[109] Peez G., Jakobs V., Wechselstromkorrosion an erdverlegten, kathodisch geschützten Rohrleitungen, 3R International, 36. Jahrgang, Heft 7/1997, Seite 325–330, 1997.

[110] Buchholz B., Koch H., Power delivery of the future, *Middle East Electricity*, Dubai, Arabic Emirates, 11/2002.

[111] Koch H., Kunze D., Monitoring system of gas insulated transmission lines (GIL), CEPSI 2006, Mumbai, India, 11/2006.

[112] Bell D., Charlson L., Holliday R., Irwin T., Lopez-Roldan J., Nixon L., High Voltage On-Site Commissioning Tests for Gas Insulated Substations Using UHF Partial Discharge Detection, IEEE, PES, Power Engineering Review, 2002.

[113] Koch H., Future aspects of gas-insulated transmission line (GIL) applications, CIGRE Colloquium, Zürich, Switzerland, 08/1999.

[114] Miyazaki A., Takinami N., Kobayashi S., *et al.*, Result of partial discharge measurements in a long-distance 275 kV GIL, *Power Engineering Review*, IEEE, PES, 2002.

[115] Luxa G.-F, Diessner A., Mosca W., Pingini A., High Voltage Testing of SF_6 Insulated Substations of Site, CIGRE Session Paris, France, 08/1986.

[116] IEC 60060-1 High-voltage test techniques – Part 1: General definitions and test requirements, IEC, Geneva, Switzerland, 2010.

[117] IEC 60270 High-voltage test techniques – partial discharge measurements, IEC, Geneva, Switzerland, 2000.

[118] Feasibility study of a North Sea network using the GIL technology to connect European off-shore wind energy projects and for the trade of electricity "GIL – TEN for Wind & Trade", 2005-E 197/05-TREN/05/TEN-E S07.63573, European Union, Commission of European Transmission Networks Electricity, Brussels, Belgium, 2005.

[119] Tenne T: Netzanschlussregeln Hoch- und Höchstspannung. Stand: 5. Oktober 2010. http://www.tennettso.de/pages/tennettso_de/Transparenz/Veroeffentlichungen/Netzanschluss/Netzanschlussregeln/tennet-NAR2010 deu.pdf

[120] Tenne T: Netzanschlussregeln Hoch- und Höchstspannung. Stand: 5. Oktober 2010. http://www.tennettso.de/pages/tennettso_de/Transparenz/Veroeffentlichungen/Netzanschluss/Netzanschlussregeln/tennet-NAR_OS_2010deu.pdf

[121] PSS®E: http://www.energy.siemens.com/hq/en/services/power-transmission-distribution/power-technologies-international/software-solutions/pss-e.htm

[122] PSS®SINCAL: http://www.energy.siemens.com/hq/en/services/power-transmission-distribution/power-technologies-international/software-solutions/pss-sincal.htm

[123] NETOMAC: http://www.energy.siemens.com/us/en/services/power-transmission-distribution/power-technologies-international/software-solutions/pss-netomac.htm

[124] Oswald B., Vergleichende Studie zu Stromübertragungstechniken im Höchstspannungsnetz, Forwind-Zentrum für Windenergieforschung, Hannover & Oldenburg, Germany, September 2005.

[125] Blaum H., Kirchesch P., Kox A., Osterholt A., Zuverlaessigkeit von 400 kV gasisolierten Leitungen (Reliability of 400 kV Gas-Insulated transmission Systems), Elektrotechnik und Automation, Germany, 1996.

[126] Blaum H., Kirchesch P., Kox A., Osterholt A., Zuverlässigkeit von 400-kV-gasisolierten Leitungen, Energietechnik, Germany, 1996.

[127] IEC 62271-203 ed1.0 High-voltage switchgear and controlgear – Part 203: Gas-insulated metal-enclosed switchgear for rated voltages above 52 kV, IEC, Geneva, Switzerland, 2003-11.

[128] C37.122 IEEE Standard for High Voltage Gas-Insulated Substations Rated Above 52 kV, IEEE, Piscataway, NJ, USA, 2010.

[129] P1677 Guide for Application and User Guide for Gas-Insulated Transmission Lines (GIL), Rated 72.5 kV and Above, IEEE, Piscataway, NJ, USA, 2010.

[130] CIGRE TB 260. N_2/SF_6 Mixtures for Gas Insulated Systems, CIGRE, Paris, France, 1996.

[131] CIGRE TB 218 Gas Insulated Transmission Lines (GIL), CIGRE, Paris, France, 2008.

[132] CIGRE TB 351 Application of Long High Capacity Gas Insulated Lines, CIGRE, Paris, France, 2009.

[133] London Infrastructure Project St John's Wood – Elstree 400 kV Cable Circuit, ABB, Mannheim, Germany, 02/2002.

[134] Gas insulated lines (GIL) – An innovative solution that respects the environment, Druckschrift, ALSTOM, Paris, France, 12/2002.

[135] Binder E., Feeberger R., Alter J., Zilavec R., Schaltanlagen und Energieausleitung im PSW Limberg II, ew Jg. 207 (2008), Heft 16.

[136] Koch H., Future needs of high power interconnections solved with Gas-Insulated Transmission Lines (GIL), PowerCon Conference, Kunming, China, 10/2002.

[137] Cooksen A. Gas Insulated Cables, IEEE Transactions on Electric Insulation, page 859-890, 10/1985.

[138] Graybill H.-W., Underground power transmission with isolated phase gas-insulated conductors, IEEE Transactions on Power Apparatus and Systems, 01/1970.

[139] Garrity T.-F., Matulic R., Rhodes G., Installation and field testing of 138 kV SF_6 gas-insulated transmission line, *IEEE Transactions on Power Apparatus and Systems, 09*-1975.

[140] Annual Report, Technische Universität Dortmund, Institute of Power Systems and Power Economics, 2008.

[141] Chakir A., Koch H., Long-term test of buried Gas Insulated Transmission Lines (GIL), CIGRE Study Committee SC15-Symposium, Dubai, 05/2001.

[142] Chakir A., Koch H., Long-term test of buried Gas Insulated Transmission Lines (GIL), IEEE PES Winter Power Meeting, New York, USA, 01/2002.

[143] Kert, C., L'Apport de Nouvelles Technologies dans l'Enfouissement des Lignies Électriques à Haute et Tres Haute Tension, Office Parlementaire d'Évaluation des Choix Sciebtifiques et Technologiques, 19. December 2001.

[144] Furtner, F. Effizientere Energiepolitik und regionale Elektrizitätsversorgung durch die EU-Elektrizitätsrichtlinie, Dissertation Katholische Universität Eichstätt-Ingolstadt, 2006.

[145] Feldmann D., Aucourt C., Boisseau C., Les càbles à isolation gazeuse, Liaison CIG, Journal Epure No. 48, Page 13-24, 12/1995.

[146] Dießner A., Koch H., Kynast E., Schuette A., Progress in high voltage testing of gas insulated transmission lines, ISH Conference, Montreal, Canada, 06/1997.

[147] Koch H., Gasisolierte Übertragungsleitungen – eine Übertragungstechnik für das 21ste Jahrhundert, ETG-Tagung, München, Germany, 10/1999.

[148] IEC 60480 Guidelines for the checking and treatment of sulphur hexafluoride (SF_6) taken from electrical equipment and specification for its re-use, IEC, Geneva, Switzerland, 2004.

[149] IEC/TR 62271-303 ed1.0 High-voltage switchgear and controlgear – Part 303: Use and handling of sulphur hexafluoride (SF_6), IEC, Geneva, Switzerland, 2008-07.

[150] Global Wind 2007 Report, Global Wind Energy Council (GWEC), 2008.

[151] Koch H., Gasisolierte Übertragungsleitungen (GIL) als Alternative zu Kabeln zur Übertragung elektrischer Energie in Ballungs- und Naturschutzgebieten, VDE-Kongress, Dresden, Germany, 10/2002.

[152] Renaud F., 220 kV Gas-Insulated Transmission Line – Palexpo Geneva Switzerland, IEEE Power Engineering Society General Meeting, Torionto, Canada,13–17 July 2003.

[153] Valov B., Lange B., Rohrig K., Heier S., Bock C., 25 GW Offshore – Windkraftleistung benötigt ein starkes Energieübertragungssystem auf der Nordsee, Universität Kassel, *Wind Kraft Journal*, Ausgabe 01/2008.

[154] Koch H., Environmental reasons for underground systems, CIGRE Session, Paris, France, 08/2002.

[155] Koch H., EMV-Planung in der Hochspannungstechnik (EMC planning for high voltage technology), ELEKTRIE, Heft 48, 1994.

[156] Benato R., Fellin L, Marzenta D., Paolucci A., Gas-Insulated Transmission Lines: excellent performance and low environmental impact, Proceedings International Symposium and Exhibition on Electric Power Engineering at the beginning of the Third Millennium, EPETM, Vol.1 pp. 385–405, Napoli, Italia, 12.–18. May 2000.

[157] Benato R., Caldon R., Mari A., Paolucci A.: Analytical formation of the state matrix of a regulated power system accounting for upfc action, Proc. of 35th Universities' Power Engineering Conference – UPEC 2000. Queens University, Belfast Ireland, pp. 5. (ACI), September 2000.

[158] VDE 0848. Safety in electric, magnetic and electromagnetic fields, DIN/VDE, Frankfurt, Germany, 2010.

[159] Connor Th., Koch H., General aspects of electromagnetic fields for high voltage systems, ISH Conference, Delft, Netherlands, 08/2003.

[160] Koch H., Offshore wind farms connected to the transmission network by GIL, Husum Wind Energy Conference Proceedings, Husum, Germany,9–13 September 2008.

[161] Sechsundzwanzigste Verordnung zur Durchführung des Bundes-Immissionsschutzgesetzes (Verordnung über elektromagnetische Felder - 26. BImSchV), Bundesministerin für Umwelt, Naturschutz und Reaktorsicherheit, 16. Dezember 1996.

[162] Verordnung über den Schutz vor nichtionisierender Strahlung (NISV), Bundesministerin für Umwelt, Naturschutz und Reaktorsicherheit, 23 December 1999.

[163] Tanizawa K., Minaguchi D., Honaga Y., Application of gas insulated transmission line in Japan, CIGRE Session Paris, France, 08/1984.

[164] HAYASHI N., SAITOU H., SHIMOHIRO D., KAWAI T.; Chubu Electric Power Co. Inc.; Nagoya, Japan; KAWAI T.; Chubu Electric Power Co. Inc.; Japan; ISHII N.; The Furukawa Electric Co. Ltd; Tokyo, Japan; HIRASAWA T.; VISCAS Corporation; Tokyo, Japan; WATANABE M., NAKAGAWA T; J-Power Systems Corp.: Japan Construction of 275 kV underground transmission line composed of continuous 2,500m long cable, Jicable Conference Proceedings, Versaille, France, 2003.

[165] Miyazaki, A.; Takinami, N.; Kobayashi, S.; Nishima, H.; Nakura, Y.; Komeda, H.; Nagano, H.; Higashi, H.; Chubu Electric Power Co.; Sumitomo Electric Industries, Ltd.: Long-Distance 275 kV GIL Monitoring System Using Fiberoptic Technology, IEEE, PES, Power Engineering Review, 08/2002.

[166] Nojima, Shimizu, Araki, Hata, Yamauchi, Installation of 275 kV 3.3 km gas-insulated transmission line for underground large capacity transmission in Japan, CIGRE Session 1998.

[167] Nojima T., Shimizu M., Miyazaki T., Study on an after-laying test for a long distance GIL, *IEEE Transactions on Power Delivery*, Vol. 13, 07/1988.

[168] Ashmore A., Electricity from the pipeline. Strom aus der Pipeline, Verband Schweizerischer Elektrizitaetswerke, 1997.

[169] The longest gas insulated line installed in the world, ALSTOM, Brochure, 06/2002.

[170] Neumann C., Erste erdverlegte GIL in Deutschland, Gasisolierte Leitung als Alternative zum Kabel, ew, Jg. 108, Heft 16, 2009.

[171] Siemens AG, ILF Beratende Ingenieure, ForWind, Grid Connection of Offshore Wind Farms, A Feasibility Study on the Application of Power Transmission Pipelines (PTP®), Brochure, Siemens, Energy, Transmission, Erlangen, Germany, 2009.

[172] Koch H., GIL – power transmission solution of the future, AEIC Conference 2006, Napa Valley, CA, USA, 09/2006.

[173] Koch H., The use of traffic tunnel for electrical power transmission, PowerGrid Europe Proceedings, Madrid, Spain, 06/2007.

[174] Koch H., Broader applications of GIL, CIGRE Session, Paris, France, 08/2002.

[175] ForWind, das Zentrum für Windenergieforschung der Universitäten Oldenburg. http://www.forwind.de/forwind/index.php

[176] Benato R., Carlini E.-M., Di Mario C., Fellin L., Paolucci A., Turri R., Gas-Insulated Transmission Lines (GIL) in railway galleries, IEEE PES Power Tech Conference, St. Petersburg, Russia, 27–30. June 2005.

[177] Benato R., Brunello P., Carlini M., Di Mario C., Fellin L., Knollseisen G., Laußegger M., Muhr M., Paolucci A., Stroppa W., Wörle H., Woschitz R., Italy–Austria GIL in the new planned railway galleries Fortezza–Innsbruck under Brenner Pass, Elektrotechnik & Informationstechnik, Vol.: 123, pages: 551–558, 2006.

[178] Koch H., Preferential Subject C2 PS2 Q2.5: Difficulties of adding new lines and expanding the transmission system, CIGRE Session, Paris, France, 08/2004.

[179] Koch H., Preferential Subject C3 PS2 Q2.8: Trading low-carbon kWh certificates, to decrease greenhouse gas emissions, impact the electric power industry, in general, and the transmission activities, CIGRE Session, Paris, France, 08/2004.

[180] Anders, G.-J., Rating of Electric Power Cables: Ampacity Computations for Transmission, Distribution and Industrial Applications, McGraw-Hill, Professional, 1997, 428 Pages.

[181] Miller D., London Infrastructure Project, Electra No. 206, 02/2003.

[182] Hänisch L., Hecklau D., Schroth R., Errichtung des 380-kV-Hochleistungskabelsystems in Berlin für den Verbundanschluß der Bewag, Elektrizitätswirtschaft, Germany, 07/1995.

[183] Henningsen, C.G., 380-kV-Diagonale durch Berlin, Elektrizitätswirtschaft, Heft $\frac{1}{2}$, Seite. 42–43, 1997.

[184] Genios Fachpresse Archive VDI Nachrichten, 380kV Kunststoffkabel soll die Versorgungssicherheit der Bundeshauptstadt erhöhen, Germany, 10. 01.1997.

[185] Henningsen C.-G., Polster K., Obst D., Berlin creates 380-kV connection with Europe, *Transmission and Distribution World*, 07/1998.

[186] Henningsen C.-G., Polster K., Obst D., Berlin's interconnection with Europe's power grid, the 380 kV diagonal link: A technical report, 50Hertz Transmission GmbH, Germany, 02/2010.

[187] Keussen U., Schneller Ch., Winter W., Ergebnisse und Konsequenzen der "dena-Netzstudie", *Energiewirtschaftl. Tagesfragen*, 55. Jg., Heft 4, 2005.

[188] Koch H., Gas Insulated Transmission Line (GIL) – Application to connect offshore wind parks to the European network, The 8th International Power Engineering Conference, IPEC 2007, Singapore, 12/2007.

[189] ForWind, das Zentrum für Windenergieforschung der Universitäten Hannover. http://www.forwind.de/forwind/index.php

[190] ILF Beratende Ingenieure, München. http://www.ilf.com/index.php?L=0

[191] Siemens Energy Transmission, Erlangen. http://www.energy.siemens.com/entry/energy/hq/en/#428370

[192] Trans-European energy networks (TEN-E). http://ec.europa.eu/energy/infrastructure/tent_e/ten_e_en.htm

[193] Bundesministerium für Wirtschaft und Technology (BMWi). http://www.bmwi.de/

[194] Bird Protection Directive (V-RL: 79/409/EWG). http://ec.europa.eu/environment/nature/legislation/birds directive/index_en.htm

[195] Boeck W., Long-term performance of SF$_6$ insulated systems by Task Force 15.03.07 of Working Group 15.03 on behalf of Study Committee 15, CIGRE Electra Publication, 2002.

[196] http://www.ptd.siemens.de/artikel0608.pdf, Guizhou-Guangdong II August 2011.

[197] The Friends of the Supergrid. http://www.friendsofthesupergrid.eu/ August 2011.

[198] The European Wind Energy Assosiation (EWEA), Annual Report 2010, http://www.ewea.org/index.php?id=178 June 2011.

[199] ISO 14040 Environmental management – Life cycle assessment – Principles and framework. ISO, Geneva, Switzerland, 2006.

[200] The European Wind Energy Assosiation (EWEA), EU Energy Policy to 2050, Achieving 80-95% emissions reductions, http://www.ewea.org/index.php?id=178 March 2011.

[201] Koch H., Einsatzreife erreicht, Siemens Report, Erlangen, Germany, 3/1999.

[202] Monard D., Diplomarbeit Nr. 1820: Comparative life cycle assessment of transmission systems for offshore wind farms, TU Darmstadt, 05/2008.

[203] Koch H., Energy transmission system with advanced technologies, World Energy Transmission Systems (WETS), Workshop Paris, France, 06/2003 http://www.see.asso.fr/jicable/wets03/pdf/wets03-1-07.pdf

[204] Thuries E., Pham V.-D., Roussel P., 420 kV three-phase compressed nitrogen insulated cable, Jicable, Versailles, France, 1995.

[205] Lequeu T., Le réseau EDF à l'épreuve des tempêtes, REE No. 2, Paris, France, 02/2000 http://www.thierry-lequeu.fr/data/REVUE 115.HTM

[206] GIL – High Power Transmission Technology, Siemens, Brochure, Erlangen, Germany, 2010, http://www.energy.siemens.com/hq/pool/hq/power-transmission/gas-insulated-transmission-lines/GIL_e.pdf.

[207] Miyazaki A., Yagi M., Kobayashi S., Min C., Hirotsu K., Nishima H., Development of partial discharge automated locating systme for power cable, IEEE, PES, Conference on Electrical Insulation and Dielectric Phenomena, Annual Report, 1998.

[208] Okuma A., Yoshida E., Takahashi T., Hoshino T., Hikata M., Miyazaki A., Partial discharge measurement in long distance SF$_6$ gas insulated transmission line (GIL), ISH, 1997.

[209] Pittroff M., Separation of SF$_6$/N$_2$ mixtures, 2nd European Conference on Industrial Electrical Equipment and Environment, pp. 61–68, Paris, France, 01/2000.

[210] Benato R., Di Mario C., Koch H. High capability applications of Long Gas Insulated Lines in Structures, IEEE PES Transactions. on Power Delivery, Vol. 22, Issue 1, pp.619–626. January 2007.

[211] Benato R., Paolucci A., Operating capability of ac EHV mixed lines with overhead and cables links, Electric Power System Research, Vol. 78/4, pp. 584–594, April 2008.

[212] Benato R., Napolitano D., Reliability Assessment of EHV Gas Insulated Transmission Lines: effect of redundancies, IEEE Transaction on Power Delivery, Vol. 23, Issue 4, pp. 2174–2181, October 2008.

[213] Lundgaard L., Particles in GIS characterization from acoustic signatures, *IEEE Transactions on Dielectrics and Electrical Insulation*, 12/2001.

[214] Hoshino T., Koyama H., Maruyama S., Hanai M., Comparison of Sensitivity Between UHF Method and IEC 60270 for Onsite Calibration in Various GIS, IEEE, PES, Transaction on Power Delivery, October 2006.

[215] Metwally I.A., Status review on partial discharge measurement techniques in gas-insulated switchgear/lines, *Electric Power Systems Research*, 04/2002.

[216] Benato R., Di Mario C., Lorenzoni A., Lines versus Cables: Consider All Factors, Transmission & Distribution World, Volume 59, Issue 11, (ISSN 1087-0849), pp. 26–32, November 2007.

[217] Cookson A.-H., Garrity T.-F., Samm R., Research and development in the United States on three-conductor and UHV rated gas insulated transmission lines for heavy load transmission, Study Committee SC 21-09, CIGRE Session Paris, France, 1978.

[218] Luxa G.-F., Boeck W., Hiesinger H., Schlicht D., Kynast E., Wiegart N. Ulrich L., Pigini A. Bargigia A.,
       Recent research activity on the dielectric performance of SF6, with special reference to very fast transients,
       Study Committee SC 15-06, CIGRE Session Paris, France, 1988.
[219] Schoeffner G., Boeck W., Graf R., Diessner A., Attenuation of UHF-signals in GIL, ISH Bangalore, India,
       08/2001.
[220] Douglas J., Costs coming down for underground, EPRI Journal, 06/1995.
[221] IEEE PES Tutorial, Gas-Insulated Substations, Substations Committee K2, http://ewh.ieee.org/cmte/
       substations/index.htm August 2011.
[222] IEC 60071-5 Insulation co-ordination - Part 5: Procedures for high-voltage direct current (HVDC) converter
       stations, 2002-06-18.
[223] Cheung K., Cheung S., Navin de Silvia R.G., Juvonen M., Singh R., Woo J.J., Compressed Air Energy
       Storage (CAES), http://www.doc.ic.ac.uk/~matti/ise2grp/energystorage_report/node7.html; Imperial College
       London, ISE2 2002/2003.
[224] Cheung C.N., Heil F., Kobayashi S., Kopejtkova D., Molony T., O'Connell P., Skyberg J.P., Taillebois I.,
       Welch, I., Report on the second international survey on high voltage gas insulated Substations (GIS), CIGRE,
       TB 150, WG 23.02, Paris, 2000-02.
[225] United Nations, World Economy Social Survey, Department of Economical and Social Affairs of the United
       Nations Secretariat, ISBN 978-92-1-109163 2 New York, 2011.
[226] Neumann C., Gas-Insulated Lines Provide EHV Solution, Transmissio & Distribution World, 2010-02-01.
[227] Siemens, Energy News, 2011-08-15 http://www.energy.siemens.com/hq/de/stromuebertragung/gasisolierte-
       uebertragungsleitungen.htm#content=Beschreibung%20.

Index

Gas-Insulated Transmission Lines (GIL), First Edition. Hermann Koch.
© 2012 John Wiley & Sons, Ltd. Published 2012 by John Wiley & Sons, Ltd.